Principles and problems in physical chemistry for biochemists

NICHOLAS C. PRICE
SENIOR LECTURER IN BIOCHEMISTRY, UNIVERSITY OF STIRLING
AND
RAYMOND A. DWEK
LECTURER IN BIOCHEMISTRY, UNIVERSITY OF OXFORD,
AND FELLOW OF EXETER COLLEGE, OXFORD

SECOND EDITION

CLARENDON PRESS · OXFORD

Oxford University Press, Walton Street, Oxford OX2 6DP

OXFORD LONDON GLASGOW
NEW YORK TORONTO MELBOURNE AUCKLAND
KUALA LUMPUR SINGAPORE HONG KONG TOKYO
DELHI BOMBAY CALCUTTA MADRAS KARACHI
NAIROBI DAR ES SALAAM CAPE TOWN

and associated companies in
BEIRUT BERLIN IBADAN MEXICO CITY NICOSIA

OXFORD is trademark of Oxford University Press

© OXFORD UNIVERSITY PRESS 1974, 1979, 1982

FIRST EDITION 1974
SECOND EDITION 1979
Reprinted with corrections 1982
Reprinted 1983

British Library Cataloguing in Publication Data

Price, Nicholas C
Principles and problems in physical chemistry for biochemists. — 2nd ed.
  1. Chemistry, Physical and theoretical
  I. Title II. Dwek, Raymond A
541'.3'024574 0D453.2   79-40341

ISBN 0-19-855511-3
ISBN 0-19-855512-1 Pbk

PRINTED IN HONG KONG

Preface to the Second Edition

IN preparing the second edition of this book we have drawn on the experience of those using the first edition, and we are grateful for all the helpful suggestions and comments which have been made. A number of new problems have been included and we have made alterations to every chapter. The major changes have been to the chapter on binding studies (where an account of co-operative behaviour has been included), to the chapter on electrochemical cells (where a section on oxidative phosphorylation has been added), to the chapter on chemical kinetics (where greater emphasis is placed on the transition-state theory), and to the chapter on enzyme kinetics (which has been expanded to include a discussion of two-substrate reactions). A new chapter on macromolecules includes some of the previous material but a new section on the ultracentrifuge has been added.

In this edition we have changed to SI units as far as possible. While we realize that some of these units may not be familiar to some teachers, we feel that we should follow the advice of bodies such as the Biochemical Society who recommend that SI units should be used. A note on these units is included.

August, 1978   NICHOLAS C. PRICE
Stirling and Oxford   RAYMOND A. DWEK

Note to the 1982 reprint

We have a number of changes to the text, notably in Chapter 7 where we have attempted to make the treatment of acid/base chemistry thermodynamically rigorous. Nearly all the changes in the text were inspired by Charles Eliot and we thank him for his enthusiasm.

N.C.P.
R.A.D.

Preface to the First Edition

THERE is a widespread belief, to which we subscribe, that the teaching of physical biochemistry is best accomplished by the students solving problems. However all too frequently the examples chosen, while being ideal for chemistry students, have little relevance for the student of biochemistry who often has difficulty in seeing the application to his subject. We have tried to set problems which we think ought to be within the capabilities of a first or second year student, and which illustrate some of the more important present-day ideas and methods. The emphasis is on principles rather than any sophisticated mathematical manipulation (unfortunately all too common in many physical chemistry textbooks). In this light we have chosen our text specifically to provide the background for the problems at the end of each chapter, and the worked examples form an integral and important part of the text—serving to illustrate several new points. The emphasis in the treatment in the book is on equilibria and rates for we believe that it is an understanding of these phenomena that provides a secure basis for more advanced topics (such as the physical chemistry of macromolecules) to be developed.

In the first part of the book we have attempted to emphasize the universal applicability of thermodynamic equations to systems in equilibrium, not only in dealing with reactions but also with properties of solutions, acids and bases and oxidation–reduction processes. The second part of the book deals with the rates of reactions, and here it is shown how many of the basic principles of chemical kinetics can be carried over into the kinetics of enzyme-catalysed reactions. For completeness we have included two short chapters on spectrophotometry and the uses of isotopes because these are of considerable importance in biochemistry. The section dealing with solution to problems contains not only the numerical answers but sufficient comment and working to enable individual students to see if, or where, they may have made errors or misunderstood certain principles. This should enable the students to work through the problems, to a large extent, on their own. Where appropriate we have tried to suggest the importance of the results. There are several appendices containing material which can be omitted at a first reading without creating difficulties in understanding the text.

Finally, it should be noted that biological systems are in general much more complicated than those dealt with by chemists. It is therefore often necessary to make drastic simplifications to perform any calculations from fundamental principles, at least at the level of this book.

November, 1973
Oxford

NICHOLAS C. PRICE
RAYMOND A. DWEK

Acknowledgements

IN writing this text we have drawn freely on the help and advice of our colleagues in the Department. In particular we should like to thank Drs. Keith Dalziel, David Brooks, John Griffiths and Simon van Heyningen. Professor H. Gutfreund made some useful criticisms of the manuscript.

We are also indebted to Mrs. Shirley Greenslade who carefully and patiently typed the original manuscript.

Note for second edition

WE should like to thank those friends and colleagues who have given help and advice in the preparation of this edition and in particular Drs. Simon Easterbrook-Smith, Lewis Stevens, Stuart Ferguson, and Peter Zavodszky.

Contents

A note on units

IN this edition we have tried, wherever possible, to use SI units. These are based on the metre–kilogramme–second system of measurement. Multiples of the basic units by powers of 10 are as follows:

Prefix		Abbreviation	Prefix		Abbreviation
$10\,(10^1)$	deca	da	10^{-1}	deci	d
10^2	hecto	h	10^{-2}	centi	c
10^3	kilo	k	10^{-3}	milli	m
10^6	mega	M	10^{-6}	micro	μ
10^9	giga	G	10^{-9}	nano	n
10^{12}	tera	T	10^{-12}	pico	p

We do not use compound prefixes, thus 10^{-9} metre (m) = 1 nm, not 1 mμm.

SI units for the various physical quantities mentioned in this edition are listed below:

Quantity	SI unit
Time	*second* (s).
Length	*metre* (m). (Supplementary units retained for convenience are dm, cm.)
Mass	*kilogramme* (kg). Note multiples are based on 1 gramme (g), i.e. mg, μg rather than μkg, nkg (see above compound prefix rule).
Volume (given in units of length cubed)	*cubic metre* (m^3). For convenience the litre (1 l = 1 dm^3) and millilitre (1 ml = 1 cm^3) are retained.
Amount of substance	*mole* (mol). This quantity contains 1 Avagadro number of basic units (e.g. electrons, atoms, or molecules).
Concentration	*mol dm*$^{-3}$ used instead of molar (M) *mol kg*$^{-1}$ used instead of molal (m)
Temperature	*degree kelvin* (K) note 0° centigrade = 273·15 K
Force	*newton* (N) (1 m kg s^{-2})
Pressure	*pascal* (Pa) (1 Nm^{-2}). *In this text however we have retained the atmosphere unit* (atm) *since the definition of standard states is then less cumbersome, being under a pressure of* 1 atm *rather than* 101·325 kPa (1 atm = 760 mm Hg = 101·325 k Pa).
Energy	*joule* (J) (1 m^2 kg s^{-2}). One calorie = 4·18 J. The gas constant, $R = 8\cdot31$ J K^{-1} mol^{-1}.

Quantity	SI unit
Electric charge	*coulomb* (C) (1 ampere second). Faraday constant = 96 500 C mol^{-1} 1 electron volt = 96·5 kJ mol^{-1}
Frequency	*hertz* (Hz) (1 s^{-1})
Viscosity	the units are kg m^{-1} s^{-1} (Note 1 centipoise = 10^{-3} kg m^{-1} s^{-1}.)

The implication of these units for the topics discussed in this book are as follows:

(i) Temperatures are quoted in degrees kelvin (or absolute). Thus 25°C is 298·15 K (in practice, this is given as 298 K).

(ii) Enthalpy, internal energy, and free energy changes are given in J mol^{-1}.

(iii) Entropies are quoted in JK^{-1} mol^{-1}.

(iv) Molar concentrations are given in terms of mol dm^{-3} rather than M.

(v) Enzyme activities are quoted in terms of katals (the amount of enzyme catalysing the transformation of 1 mol substrate per second). Specific activities are quoted in terms of katal kg^{-1}.

(vi) The curie (Ci) is redundant as a unit of radioactivity; 1 Ci = $3·7 \times 10^{10}$ s^{-1}. (We have retained this unit, however, in this text.)

For fuller discussions SI units the following may be consulted:

Quantities, units, and symbols (2nd edn). The Royal Society, London (1975). *Physicochemical quantities and units* (2nd edn). M. L. McGlashan, Royal Institute of Chemistry, London (1971).

1 The first law of thermodynamics

What is thermodynamics?

THERMODYNAMICS is concerned with the bulk behaviour of substances. Certain empirical laws are used to derive equations which can then be used to calculate the final results of processes. It does not deal with the rates of such processes; this aspect is encompassed in the subject of kinetics. To the biochemist, the importance of thermodynamics lies in its ability to predict the position of equilibrium in a system. For example, we might wish to study the equilibrium position of, and energy changes involved in, the reaction:

$$ATP + H_2O \rightarrow ADP + phosphate.$$

As we shall see later, determination of the energy changes would involve measuring the concentrations of the various substances taking part in the reaction when the system has come to equilibrium. In this system, however, the equilibrium concentration of ATP is too small to be accurately measured.

However, we can make measurements of the equilibrium concentrations of reactants in the reactions:

(1) $Glutamate + NH_4^+ + ATP \rightarrow Glutamine + ADP + phosphate$

and

(2) $Glutamine + H_2O \rightarrow Glutamate + NH_4^+.$

Using thermodynamics we could then predict the position of equilibrium and calculate the energy changes involved in the reaction of interest, i.e.

$$ATP + H_2O \rightarrow ADP + phosphate$$

(which is the sum of reactions (1) and (2) above).

The position of equilibrium in a biochemical system is often of crucial importance. Many processes are regulated by the binding of one molecule to another. For instance, the catalytic activity of many enzymes is affected by the binding of small molecules:

$$Enzyme + R \rightleftharpoons (Enzyme. R).$$

$$\begin{pmatrix} active \\ form \end{pmatrix} \quad \begin{pmatrix} regulator \\ molecule \end{pmatrix} \quad \begin{pmatrix} inactive \\ form \end{pmatrix}$$

An example of such a system might be, for instance, pyruvate dehydrogenase which can be inhibited by the binding of acetyl CoA (which is in fact a product of the reaction).

Knowledge of the thermodynamic quantities involved in this reaction allows us to predict the position of equilibrium under any specified set of conditions (concentration of reactants, temperature, etc.) and hence to calculate the amount of enzyme in the active form under these conditions.

Some basic definitions

A *system* consists of matter which is capable of undergoing a change. A complete definition includes the matter contained and the pressure, volume, and temperature.

The *surroundings* are anything in contact with the system which can influence its state.

A *process* is any change which is taking place in the system.

Systems can interact with their surroundings via a flow of heat or work, or matter. Systems such as the cell which interact via a flow of matter are called *open* systems but initially we will consider *closed* systems where only heat and work changes occur. If two systems, or a system and its surroundings, are at the same temperature they are said to be in thermal equilibrium.

Statement of the First Law of Thermodynamics

The First Law is the law of conservation of energy and defines a quantity known as *internal energy*. If a system is isolated (that is it does not interact with its surroundings) the First Law states that its energy will remain constant irrespective of any processes taking place. Suppose however the system interacts with its surroundings. The First Law states that 'a change in internal energy ΔU occurs such that

$$\boxed{\Delta U = \Delta q - \Delta w}\,.$$

Δq is the heat *absorbed* by the system during the interaction and Δw is the work done *on* the system *by the* its surroundings.' Any surplus heat energy is gained as internal energy so no energy is destroyed or created. We should note that the change in internal energy is independent of the path taken to go from the initial to the final state:[†]

$$\Delta U = U_f - U_i.$$

[†] Properties of a system which depend only on the initial and final states of the system are termed *state functions*. Examples that we shall meet include internal energy, enthalpy, entropy, and Gibbs free energy.

If this were not so, a cyclic process could create or destroy energy and perpetual motion would be possible.

The symbols Δ in the above equations refer to large measurable changes in the quantities U, q, w. We could, of course, also express the First Law in terms of very small changes in these quantities:-

$$\delta U = \delta q + \delta w.$$

Applications of the First Law of Thermodynamics

In biochemical systems several forms of work are carried out. Some examples are: the mechanical work performed by muscles, the electrical work required to charge nerve membranes, and the chemical work in synthesis of large molecules, or to produce the light emitted by glow worms. The First Law of Thermodynamics can be applied to each of these processes, provided the appropriate Δw (work) term is used. However initially we will consider only the work done against a pressure when chemical reactions proceed.

Consider a small change (δV) in volume when such a reaction proceeds. The work done (δw) against the pressure is given by

$$\delta w = -P\,\delta V.$$

(Note that if δV is small enough, the pressure P can be considered as effectively constant.)

Now from the First Law

$$\delta U = \delta q - P\,\delta V.$$

For large measurable changes

$$\Delta U = \Delta q - \int P\,\mathrm{d}V.$$

The pressure must be known as a function of volume before the work term (Δw) can be calculated by integration.

Reactions are usually studied at constant volume or pressure to simplify the integral above. At *constant volume* $\mathrm{d}V$ will be zero, so that $\Delta U = \Delta q_\mathrm{v}$, while at *constant pressure*

$$\int_{V_i}^{V_f} P\,\mathrm{d}V = P(V_f - V_i) = P\,\Delta V$$

and so

$$\Delta U = \Delta q_\mathrm{p} - P\,\Delta V.$$

Clearly measuring Δq_v, the heat absorbed at constant volume, e.g. in a bomb calorimeter, gives ΔU for the reaction directly. From the above equation however, Δq_p, the heat absorbed at constant pressure, equals $\Delta U + P\Delta V$. For convenience we will define an energy function H (known as *enthalpy*) such that

$$\boxed{H = U + PV}\;.$$

Then at constant pressure,

$$(H_f - H_i) = (U_f - U_i) + P(V_f - V_i)$$

or

$$\Delta H = \Delta U + P\,\Delta V$$

for any reaction.

Consequently ΔH (the change in enthalpy of the reaction) equals Δq_p. Since most chemical reactions are studied at constant pressure, enthalpy changes (ΔH) rather than internal energy changes (ΔU) are usually quoted. For ideal gases a simple relationship between ΔH and ΔU exists. Since $PV = nRT$, then for a reaction at constant temperature and pressure, which involves a change Δn in the number of moles of gas, we have

$$P\Delta V = \Delta n (RT)$$

$$\boxed{\Delta H = \Delta U + \Delta n (RT)}\;.$$

For reactions in solution, the volume changes are normally negligible and so $\Delta H = \Delta U$.

Worked examples

(1) The oxidation of solid lactic acid was studied in a bomb calorimeter at 291 K. The heat released was 1367 kJ mol^{-1}: find ΔH for the process.

Solution

Now Δq_v is the heat *absorbed* at constant volume and so

$$\Delta U = \Delta q_v = -1367 \text{ kJ mol}^{-1}$$

$$CH_3 \cdot CHOH \cdot CO_2H(s) + 3O_2(g) \;\rightarrow\; 3CO_2(g) + 3H_2O(l).$$

Δn equals the change in the number of moles of gaseous components and here Δn is zero. So ΔH is also $\underline{-1367 \text{ kJ mol}^{-1}}$ for this reaction.

(2) In a bomb calorimeter, the combustion of fumaric acid released 1330 kJ mol^{-1}; calculate ΔH for this process ($T = 291$ K). $R = 8\cdot31$ J K^{-1} mol^{-1}.

Solution

$$\begin{array}{c}
\text{H} \qquad\qquad \text{CO}_2\text{H} \\
\diagdown \qquad\qquad \diagup \\
\text{C}=\text{C} \\
\diagup \qquad\qquad \diagdown \\
\text{HO}_2\text{C} \qquad\qquad \text{H}
\end{array} \quad (s)+3O_2(g) \rightarrow 4CO_2(g)+2H_2O(l).$$

Here Δn is $+1$, so

$$\Delta H = -1330 + \frac{(1)(8\cdot31)(291)}{1000} \text{ kJ mol}^{-1}$$

$$= \underline{-1327\cdot6 \text{ kJ mol}^{-1}}.$$

These examples show that the complete oxidation (combustion) of an organic compound leads to the release of a large amount of energy.† For instance in the cell, much of the energy used in metabolism is obtained via the oxidation of pyruvate to carbon dioxide in the tricarboxylic acid cycle. In this case the energy released is utilized in a series of steps, resulting in the conversion of 15 moles of ADP to ATP per mole of pyruvate oxidized.

Thermochemistry

Thermochemistry is the practical application of the First Law to chemical reactions. For convenience several different types of heat of reaction are defined.

The *enthalpy of reaction* is defined as the heat *absorbed* during a reaction at constant pressure. Consequently ΔH is always negative for exothermic reactions (i.e. reactions in which heat is evolved).

The *enthalpy of combustion* is the heat *absorbed* at constant pressure when one mole of substance is converted to its combustion products, e.g. CO_2, H_2O etc.

An important quantity is the *enthalpy of formation* (ΔH_f) of a compound. This is the heat absorbed at constant pressure when a compound is formed from its elements in their most stable form. The value of ΔH_f at 298 K at 1 atm. pressure is known as the *standard enthalpy of formation* (denoted by ΔH_f°). This quantity is often difficult to measure directly but this difficulty can be overcome by making use of *Hess's Law of Constant Heat Summation*. This is a restatement of the First Law and says that ΔH for any chemical reaction is independent of the reaction path (i.e. enthalpy is a *state function* and depends only on the initial and final states). For example, if one requires ΔH_f° for glucose, the following reactions can be considered:

(1) $C(\text{graphite}) + O_2(g) \rightarrow CO_2(g) \qquad \Delta H_1^\circ = -393\cdot1 \text{ kJ mol}^{-1}$

† This amount of energy is large compared with that of most biochemical reactions (see Chapter 3).

(2) $H_2(g) + \frac{1}{2}O_2(g) \rightarrow H_2O(l)$ $\Delta H_2^\circ = -285 \cdot 5 \text{ kJ mol}^{-1}$

(3) $C_6H_{12}O_6(s) + 6O_2(g) \rightarrow 6CO_2(g) + 6H_2O(l)$
$$\Delta H_3^\circ = -2821 \cdot 5 \text{ kJ mol}^{-1}.$$

We are interested in the reaction

$$6C(\text{graphite}) + 6H_2(g) + 3O_2(g) \rightarrow C_6H_{12}O_6(s)(\Delta H_f^\circ).$$

From Hess's Law, $\Delta H_f^\circ = 6\Delta H_1^\circ + 6\Delta H_2^\circ - \Delta H_3^\circ = -1250 \cdot 1 \text{ kJ mol}^{-1}$. So glucose is an *exothermic*† compound (i.e. ΔH_f° is negative). An *endothermic* compound (such as ethene) has a positive value of ΔH_f° (in this case $54 \cdot 3 \text{ kJ mol}^{-1}$).

PROBLEMS

1. The heats of combustion of fumaric and maleic acids are $-1335 \cdot 9$ and $-1359 \cdot 1 \text{ kJ mol}^{-1}$ respectively at 298 K. Calculate the heat of formation of each of these isomers from its elements and also the heat of isomerization, i.e. maleic $\rightarrow$ fumaric at 298 K. (Heats of formation of CO_2 and H_2O are $-393 \cdot 1$ and $-285 \cdot 5 \text{ kJ mol}^{-1}$ respectively at 298 K.)

2. The values for ΔU on combustion of glucose and stearic acid are -2880 and $-11\ 360 \text{ kJ mol}^{-1}$ respectively (at 310 K). Evaluate the ΔH for each of these processes. (Glucose is $C_6H_{12}O_6$ and stearic acid is $C_{18}H_{36}O_2$: both are solids.) $R = 8 \cdot 31 \text{ J K}^{-1} \text{mol}^{-1}$.

   In the light of your results comment on the relative suitability of glycogen (a polymer made up of glucose units) and fatty acids as energy reserves in the body.

3. The heats of combustion of glucose and pyruvic acid at 298 K and 1 atm pressure are $-2821 \cdot 5$ and $-1170 \cdot 4 \text{ kJ mol}^{-1}$ respectively. Calculate the enthalpy change for the conversion of glucose to pyruvic acid in the reaction:
$$\text{Glucose} + O_2 \rightarrow 2 \text{ pyruvic acid} + 2H_2O.$$

4. On arousal from hibernation, a hamster can raise its body temperature by as much as 30 K. Assuming that the heat required for this arises from the combustion of fatty acids calculate the weight of fatty acid which would need to be oxidized in order to warm up a 100 g hamster. (Use the data provided in question 2 for the combustion of stearic acid and assume that the specific heat of the hamster tissue is $3 \cdot 30 \text{ J K}^{-1} \text{g}^{-1}$.)

5. (a) Using Hess's Law calculate the lattice energy of $MgCl_2$, i.e. ΔH for the process
$$MgCl_2(s) \rightarrow Mg^{2+}(g) + 2Cl^-(g),$$

† ΔH_f° is the enthalpy of the products *minus* that of the reactants. Thus a negative value of ΔH_f° indicates that the system has lost enthalpy (to the surroundings) and vice versa. This *sign convention* applies throughout this book.

given the following data:

$$Mg(s) \rightarrow Mg(g) \qquad \Delta H = 167 \cdot 2 \text{ kJ mol}^{-1}$$
$$Mg(g) \rightarrow Mg^{2+}(g) \qquad \Delta H = 2182 \cdot 0 \text{ kJ mol}^{-1}$$
$$Cl_2(g) \rightarrow 2Cl.(g) \qquad \Delta H = 241 \cdot 6 \text{ kJ mol}^{-1}$$
$$Cl.(g) \rightarrow Cl^-(g) \qquad \Delta H = -364 \cdot 9 \text{ kJ mol}^{-1}$$

and that the heat of formation of $MgCl_2$ is $-639 \cdot 5$ kJ mol^{-1}.

(b) The heat of solution of anhydrous solid $MgCl_2$ is $-150 \cdot 5$ kJ mol^{-1}. Using the results from part (a) above, calculate the heat of hydration of the gaseous ions.

(c) If a value of $-383 \cdot 7$ kJ mol^{-1} is assumed for the heat of hydration of a gaseous Cl^- ion, what is the heat of hydration for gaseous Mg^{2+}? The corresponding value for gaseous Ca^{2+} is -1560 kJ mol^{-1}. Comment.

6. One of the uses of thermochemical measurements is in the determination of 'bond energies'. The C–H bond energy in methane for example would be one quarter of the enthalpy change of the reaction

$$CH_4(g) \rightarrow C(g) + 4H(g).$$

Determine this bond energy given the following data:

$$C(s) + 2H_2(g) \rightarrow CH_4(g) \qquad \Delta H_1^\circ = -74 \cdot 8 \text{ kJ mol}^{-1}$$
$$H_2(g) \rightarrow 2H.(g) \qquad \Delta H_2^\circ = +434 \cdot 7 \text{ kJ mol}^{-1}$$
$$C(s) \rightarrow C(g) \qquad \Delta H_3^\circ = +719 \cdot 0 \text{ kJ mol}^{-1}.$$

Assuming that the C–H bond energy is the same in methane and ethene and that the ΔH_f° for ethene is $+54 \cdot 3$ kJ mol^{-1}, calculate the C$=$C bond energy in ethene.

7. The ΔU for hydrogenation of ethene (to give ethane) is $-133 \cdot 8$ kJ mol^{-1} (at 298 K). Calculate the value of ΔH for this process. The value of ΔH for the hydrogenation of benzene is $-208 \cdot 2$ kJ mol^{-1} (at 298 K). Compare these results and work out by how much benzene is more stable than a hydrocarbon containing three isolated double bonds. (This quantity is known as the resonance energy of benzene.)

2. The second law of thermodynamics

Introduction

THE First Law of thermodynamics deals simply with the energy balance in a given process or reaction. According to this law a process (reaction) can occur in either direction, provided that the internal energy of the system remains constant. However, it is found in practice that many processes will only occur in one direction. For instance, heat will flow from a system at one temperature to another system at a lower temperature, but the reverse flow does not occur. It is observed that a beaker of water will not suddenly divide its internal energy so that half the water freezes while the other half heats up (or boils!) (although from the First Law there is nothing to suggest that this cannot happen). Clearly we need some criterion to tell us about the likelihood of a process occurring in a certain direction. This essentially is the subject of the second law of thermodynamics.

However before we discuss the second law we must introduce a new concept—that of the entropy of a system.

The meaning of entropy

The simplest way to introduce entropy is to think of it as measuring the degree of disorder (or randomness) of a system. Thus we find that the entropy of water (at 273·15 K) is considerably greater than that of ice (at 273·15 K), reflecting the more ordered arrangement of the molecules in the ice structure. Similarly the entropy of water vapour at 373·15 K is greater than that of the liquid at this temperature, again reflecting the greater freedom of motion (randomness) of the molecules in the gaseous phase. If we consider the melting of ice, at 273·15 K, for example, we achieve the increase in entropy (or disorder) by the input of heat energy (known as the *latent heat* of melting or fusion). It follows that entropy and heat changes must be related in some way.

Statement of the Second Law

The Second Law relates the change in the *entropy* of a system (dS) during some process (or reaction) to the heat absorbed by the system (dq) at a temperature, $T(K)$. The law can be expressed as

$$dS \geq dq/T.$$

The equality in the above definition (i.e. when $dS = dq/T$) refers to a process which is *reversible* whereas the inequality refers to an *irreversible* process.[†]

Reversible and irreversible processes

A process is said to be *reversible* (or carried out *reversibly*) when a system can be taken from state A to state B and then back to state A, with the surroundings also being restored to their original state. *There are thus no permanent changes in either the system or its surroundings.* Reversible processes do not occur in nature but they represent the limits of certain types of processes. Consider, for example, a gas enclosed by a piston with the pressure on the two sides of the piston equal. If this equilibrium is perturbed by lowering the opposing pressure, the gas will expand; conversely the gas will contract if the opposing pressure is increased. When the changes in the opposing pressure are made infinitesimally small, the process of expansion or contraction approaches reversibility. A reversible process thus corresponds to a series of equilibrium states in which one state is converted to another by an infinitesimal change in pressure or other variable such as temperature, concentration, etc. From this it follows that reversible processes occur infinitely slowly.

Irreversible processes are those which proceed spontaneously, i.e. at a finite rate. Examples of such processes include the sudden expansion of a gas into a vacuum, the flow of heat from a hot bath to a cold bath, and the flow from matter from a region of high concentration to one of low concentration.

Most chemical reactions are carried out irreversibly, i.e. the reactants are mixed and the reactions allowed to proceed,[‡] In these situations we note from the second law that $dS > dq/T$. This inequality provides a measure of the spontaneity of, or the tendency of, the reaction to proceed in a given direction. *So long as $dS > dq/T$ the reaction will proceed.*

The use of thermodynamics would be very limited if it only allowed us to consider inequalities. Of course most processes *are* carried out irreversibly, but the assumption of reversibility allows us to derive equations which are found to hold very well in practice.

[†] Of course heat is transferred from the *surroundings* to the system and hence the entropy of the surroundings must decrease. Actually in a reversible process the entropy is conserved (i.e. the entropy lost by the surroundings equals the entropy gained by the system), while in an irreversible process the entropy gained by the system is *greater* than that lost by the surroundings (i.e. there is overall a *net* increase in entropy).

[‡] We shall see later (Chapter 8) that with electrochemical cells the reactants are mixed but the reaction is prevented by applying an opposing e.m.f.

The concept of free energy

The statement of the Second Law tells us that a reaction or a chemical process will proceed so long as $dq < T\,dS$. The vast majority of reactions are carried out at constant temperature and pressure and under such conditions

$$dq = dH$$

i.e. the heat absorbed is equal to the change in enthalpy for the process (see Chapter 1). Substituting for dq, we thus have $dH < T\,dS$ or $dH - T\,dS < 0$. So long as this condition is valid the reaction will proceed. The importance of the quantity $(dH - T\,dS)$ as a measure of the extent to which a reaction will proceed is such that it is given a special symbol (dG), i.e.

$$dG = dH - T\,dS.$$

Thus if a process occurs then $dG < 0$. The process will continue until an equilibrium is attained. At equilibrium, the second law states that

$$dq = T\,dS$$

so that under our conditions of constant temperature and pressure

$$dH = T\,dS.$$

We then have that at *equilibrium*

$$\boxed{dG = dH - T\,dS = 0}.$$

We are now in a position to make a statement about the possibility of a process occurring.

'Processes will occur at constant temperature and pressure so long as the changes in G are negative, i.e. $dG < 0$. The equilibrium condition is that $dG = 0$.' The symbol G is termed the *Gibbs free energy*.

Fig. 2.1 is a pictorial illustration of the above general statement.

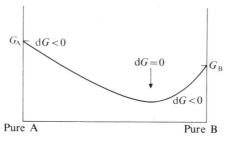

FIG. 2.1 The change in Gibbs free energy as the reaction A $\rightleftarrows$ B proceeds. At equilibrium G is a minimum (i.e. $dG = 0$).

In the same way as we refer to the enthalpy and entropy of a substance, we can refer to its Gibbs free energy. This is defined simply as:

$$G = H - TS$$

For a *reaction* at constant temperature and pressure the change in Gibbs free energy (ΔG) is given by

$$\boxed{\Delta G = \Delta H - T\,\Delta S}$$

where ΔH and ΔS are the corresponding changes in enthalpy and entropy.

From our previous discussions it is clear that a reaction will proceed if $\Delta G < 0$ and will be at equilibrium if $\Delta G = 0$. ΔG is a measure of the work (or energy) that can be obtained from a reaction. At equilibrium no work can be obtained (since there is now no tendency for the reaction to proceed).

An everyday example is afforded by a battery. Here a finite voltage is obtained only because the chemical reaction in the battery is not at equilibrium. As the reaction proceeds towards equilibrium, the voltage drops, until, at equilibrium, it becomes zero (i.e. the battery is said to be 'run down').

Like enthalpy and entropy, G is a *state function*—that is, changes in its value depend only on the initial and final states of the system. This means we can add and subtract values of ΔG for different reactions, in the same way that we did for changes in enthalpy (Chapter 1).

For most processes, ΔG is dominated by either the ΔH or the $T\,\Delta S$ term, and the process is said to be enthalpy or entropy controlled. If ΔH for a reaction is large and negative, ΔG will be negative unless ΔS is also large and negative. Consequently highly exothermic reactions (ΔH negative) are almost invariably favourable on free energy grounds (i.e. ΔG negative) and ΔH dominated. However, for some processes ΔH is very small and the entropy term dominates. Examples of such processes are metal ion complex formation with polydentate ligands such as $EDTA^{4-}$, the formation of certain compounds such as CS_2 and some cases of protein denaturation. These processes are all favourable around room temperature because the $T\,\Delta S$ term is large enough to make ΔG negative even though ΔH is positive.

If the metal ion complexing is considered, the process could be for example:

$$Mg(H_2O)_6^{2+} + EDTA^{4-} \rightarrow [Mg(EDTA)(H_2O)]^{2-} + 5H_2O$$

The value of ΔH is about $+12 \cdot 5$ kJ mol^{-1} but the entropy term is highly favourable (i.e. positive) as each $EDTA^{4-}$ ligand releases five water molecules from an ordered solvation shell. This makes the ΔG for the process negative, i.e. complex formation is favoured. Similarly if CS_2 formation is considered,

$$C(s) + 2S(s) \rightarrow CS_2(l)$$

entropy considerations always favour liquid formation from two ordered solid phases. The $T \Delta S$ term is sufficiently large to overcome the positive value of ΔH, so CS_2 can be formed at room temperature.

At 313 K and pH 3, chymotrypsin can be denatured. ΔH is highly unfavourable, 62·7 kJ mol^{-1}, but ΔS is 1839 J K^{-1} mol^{-1} at 313 K and so ΔG is -513 kJ mol^{-1}. As a result denaturation is highly favourable.†

The denaturation of a protein involves the destruction of a large number of non-covalent bonds (e.g. H-bonds) which are responsible for maintaining the highly ordered *tertiary* structure of the active form. Denaturation has thus increased the disorder (or randomness) of the system, and it is for this reason that the *denatured* form is often referred to as the *random coil* form.

Standard states

The value of G of a substance in its standard state is denoted by $G°$. The *standard state* of a substance (which is then a common reference point) *is defined as that substance under a pressure of 1 atmosphere*. This definition does not include temperature, so we can talk about the standard free energies of a compound at 298 K or 310 K or any other temperature and denote them by $G°_{298}$ or $G°_{310}$, etc.

For a reaction we can then write $\Delta G°$ for the change in free energy when reactants in their standard state are converted to products in their standard state. The superscript (°), included in the enthalpy and entropy terms, has a similar meaning.

$$\Delta G° = \Delta H° - T \Delta S°.$$

Worked examples

(1) Given that $\Delta G°$ for the hydrolysis of ATP to ADP and phosphate is $-30·5$ kJ mol^{-1} and $\Delta G°$ for the hydrolysis of ADP to AMP and phosphate is $-31·1$ kJ mol^{-1} what is the $\Delta G°$ for the process below

$$ATP + AMP \rightarrow 2ADP$$

under these conditions? This is known as a 'coupled' reaction. (The hydrolysis of ATP is linked to the phosphorylation of AMP.)

---

† The active form is actually stable because of the high 'activation energy' required to denature it (see Chapter 9).

Solution

We have

    (i)   ATP $\rightarrow$ ADP + P_i† $\Delta G° = -30\cdot5$ kJ mol^{-1}

    (ii) ADP $\rightarrow$ AMP + P_i $\Delta G° = -31\cdot1$ kJ mol^{-1}.

Since G is a *state function*, we can add or subtract the ΔG's. Hence by subtraction of (2) from (1)

$$\text{ATP} + \text{AMP} \rightarrow 2\text{ADP} \qquad \Delta G° = \underline{0\cdot6 \text{ kJ mol}^{-1}}.$$

Thus the equilibrium in this overall reaction lies slightly to the left.

(2) At 310 K, the $\Delta G°$ for ATP hydrolysis is $-30\cdot5$ kJ mol^{-1}. Calorimetric measurements give $\Delta H° = -20\cdot1$ kJ mol^{-1}. What is the value of $\Delta S°$ for this process?

Solution

  Now $\Delta G° = \Delta H° - T\,\Delta S°$

$$\therefore \quad \Delta S° = \frac{\Delta H° - \Delta G°}{T}$$

$$= \frac{-20\cdot1 + 30\cdot5}{310}(1000) \text{ J K}^{-1} \text{ mol}^{-1}$$

$$\Delta S° = \underline{33\cdot5 \text{ J K}^{-1} \text{ mol}^{-1}}.$$

The 'biochemical' standard state

In many processes of biochemical interest, there is a net uptake or release of protons as the reaction proceeds, e.g. the hydrolysis of ATP at a pH around 8 can be formulated as:

$$\text{H}_2\text{O} + \text{ATP}^{4-} \rightarrow \text{ADP}^{3-} + \text{HPO}_4^{2-} + \text{H}^+$$

Strictly speaking, $\Delta G°$ is the change in free energy when the reactants in their standard states are converted to products in their standard states. For the solutes, i.e. ATP^{4-}, ADP^{3-}, HPO_4^{2-}, and H^+ we shall see later that the standard state is defined as a one molar solution of these ions (still, of course, at 1 atmosphere pressure). For H^+ a 1 mol dm^{-3} solution would correspond to a pH = 0. This standard state is not of much interest to the biochemist who is normally concerned with reactions in solution where the pH is around neutrality ($\sim$7). For this reason, it is convenient to define a new 'biochemical' standard state, where all the components are present in

---

† P_i is often used as an abbreviation for phosphate.

their standard states *except* H^+ which is present at 10^{-7} mol dm^{-3} (pH = 7). The 'biochemical' standard free energy change is denoted by $\Delta G^{\circ\prime}$.

PROBLEMS

1. Calculate the changes in entropy for the following processes:
   (a) Ice → water (273·15 K)
   (b) Water → water vapour (373·15 K).
   The latent heat of fusion and vaporization are 334 and 2424 kJ kg^{-1} respectively.

2. Could the following reactions proceed spontaneously?
   (a) Malate^{2-} → fumarate^{2-} + H_2O (298 K)
   (b) Leucine + glycine → leucylglycine + H_2O (298 K)
   (c) Oxaloacetate^{2-} + H_2O → pyruvate$^-$ + HCO_3^- (298 K)
   (d) Ice → water (263 K).
   Values for the standard Gibbs free energies of formation at 298 K in kJ mol^{-1} are as follows:

malate^{2-}	−844	leucine	−341
fumarate^{2-}	−604	glycine	−373
oxaloacetate^{2-}	−796	leucylglycine	−464
pyruvate$^-$	−474	water	−237
HCO_3^-	−586		

3. The heat of neutralization of chloracetic acid by OH^- is −62·3 kJ mol^{-1}. The ΔG° for ionization (at 298 K) is 17·1 kJ mol^{-1}. Calculate the ΔS° for ionization of chloracetic acid. (For $H^+ + OH^- \rightarrow H_2O$; $\Delta H^\circ = -56\cdot8$ kJ mol^{-1}.)

4. The transfer of methane from an inert (non-polar) solvent to water is often used as a model for consideration of 'hydrophobic' forces. Given the following values (at 298 K):

   $$CH_4 \text{ (inert solvent)} \rightarrow CH_4(g) \qquad \Delta G^\circ = -14\cdot6 \text{ kJ mol}^{-1}$$
   $$\Delta H^\circ = +2\cdot1 \text{ kJ mol}^{-1}$$

   and

   $$CH_4(g) \rightarrow CH_4 \text{ (aqueous solution)} \qquad \Delta G^\circ = +26\cdot3 \text{ kJ mol}^{-1}$$
   $$\Delta H^\circ = -13\cdot4 \text{ kJ mol}^{-1}$$

   calculate the ΔG° and ΔH° for transfer of CH_4 from an inert solvent to water. Comment on the differences between the two quantities.

5. The binding of inhibitor to trypsin was studied. $\Delta G^{\circ\prime}$ for the formation of the complex was found to be −20·9 kJ mol^{-1} (at 298 K). Calorimetric measurements showed that the enthalpy change was zero. What is the $\Delta S^{\circ\prime}$ for the process below?

   Trypsin + Inhibitor → (Trypsin. Inhibitor complex).

   Comment on these data.

6. The enzyme glutamine synthetase catalyses the reaction

   $$\text{Glutamate} + NH_4^+ + \text{ATP} \xrightarrow{\text{Mg}^{2+}} \text{glutamine} + \text{ADP} + P_i.$$

For this reaction $\Delta G^{o\prime} = -15\cdot3$ kJ mol^{-1} at 310 K. $\Delta G^{o\prime}$ for the hydrolysis of ATP (to yield ADP and P$_i$) is $-30\cdot5$ kJ mol^{-1}. Calculate the value of $\Delta G^{o\prime}$ for the hydrolysis of glutamine to yield glutamate and NH$_4^+$.

7. The value of $\Delta G^{o\prime}$ for the hydrolysis of creatine phosphate

$$\text{Creatine phosphate} + H_2O \;\rightarrow\; \text{creatine} + P_i$$

is $-37\cdot6$ kJ mol^{-1} at 310 K. Could this reaction be used to favour the synthesis of ATP from ADP and P$_i$? ($\Delta G^{o\prime}$ for the hydrolysis of ATP is $-30\cdot5$ kJ mol^{-1} at 310 K.)

8. The enthalpy change in unfolding ribonuclease at pH 6 is $+209$ kJ mol^{-1}. If the entropy change on unfolding is $+554$ J K^{-1} mol^{-1}, calculate the free energy of unfolding at 298 K. Comment on the magnitude of the numbers given above.

3. Chemical equilibrium

Introduction

BIOCHEMICAL systems are rarely at equilibrium as matter is constantly entering and leaving cells, resulting in a *steady state* situation rather than a true equilibrium. (The term steady state is often used to describe a situation in which the concentration of a substance remains almost constant, i.e. the rate of its formation equals the rate of its breakdown.) In these circumstances equilibrium thermodynamics is not strictly applicable and non-equilibrium thermodynamics has been developed to describe such processes.† However, using equilibrium thermodynamics, a good idea of which cell reactions are favourable *in vitro* can be obtained, thus providing the biochemist with the means to understand some of the energy balances involved in cellular processes.

Equilibrium and non-equilibrium reactions

If we examine a biochemical pathway in detail we often find that the concentrations of metabolites are such that some of the individual steps are approximately at equilibrium, whereas others are maintained far away from equilibrium. It seems that the finding that a given step is not at equilibrium, constitutes good evidence that this step may be a control point in the overall pathway. Examples in the glycolytic pathway include:

$$G\text{-}6\text{-}P \;\rightarrow\; F\text{-}6\text{-}P \text{ (catalysed by phosphohexoseisomerase)}$$

which is essentially at equilibrium and

$$F\text{-}6\text{-}P + ATP \;\rightarrow\; FDP + ADP \text{ (catalysed by phosphofructokinase)}$$

which is far from equilibrium.

---

† The difference between the two types of thermodynamics can be described as follows. Equilibrium thermodynamics considers processes which are reversible (e.g. reactions at equilibrium). In these circumstances the entropy of a system increases by an amount $dS = dq/T$, where dq is the heat transferred. This entropy is lost by the *surroundings* so that the *net* change in entropy (i.e. in the system and its surroundings) equals zero. Now for irreversible processes such as systems not at equilibrium, $dS > dq/T$ and there is a *net increase of entropy* accompanying the process. Non-equilibrium thermodynamics shows that the rate of increase of entropy in the system and its surroundings is a *minimum* when the system is in a *steady state*. Thus non-equilibrium thermodynamics states that for systems away from equilibrium, the *steady state* is the *most ordered* state. To a very good approximation we can consider living cells as being in a steady state with the concentration of intermediary metabolites remaining approximately constant. It is important to note that when a reaction is at equilibrium no useful work can be obtained from it (the example of the 'run down' battery in Chapter 2), and it is only when non-equilibrium conditions prevail (such as in a steady state) that useful work can be obtained.

Phosphofructokinase is considered to be one of the key regulatory enzymes in glycolysis. For a further discussion the book by Newsholme and Start in the reading list should be consulted.

The relationship between $\Delta G°$ and equilibrium constant

In Chapter 2 we mentioned that ΔG provided a measure of the tendency of a reaction to proceed towards equilibrium (where $\Delta G = 0$). We now wish to derive an equation linking the standard free energy change for a reaction ($\Delta G°$) and the equilibrium constant of the reaction. It is then possible to *predict* the position of equilibrium in reactions, since the values of the standard free energies of reactants and products are often available in tables or can be measured.

The starting point is the relationship

$$G = H - TS,$$

which applies to each component in a system.

Now since

$$H = U + PV,$$

$$G = U + PV - TS;$$

differentiating (i.e. considering a small change)

$$dG = dU + P\,dV + V\,dP - T\,dS - S\,dT.$$

Now from the First Law, for any process

$$dU = dq - P\,dV,$$

where dq is the heat absorbed by the system and $P\,dV$ is the work done by the surroundings on the system. From the Second Law for a *reversible* process

$$dS = dq/T$$

i.e.

$$dq = T\,dS.$$

Then we have

$$\underline{dU = T\,dS - P\,dV.}$$

Substituting in the expression for dG, we obtain

$$dG = V\,dP - S\,dT.$$

At constant temperature (i.e. $dT = 0$)

$$dG = V\,dP.$$

We can integrate this equation if we know V as a function of P. For instance let us apply this equation to a perfect gas, for which $PV = nRT$

$$dG = \frac{nRT}{P}\,dP.$$

Integrating this between the standard state (where $P = P° = 1$ atm and $G = G°$) and any arbitrary value of P, then we obtain

$$\int_{G°}^{G} dG = \int_{P°}^{P} \frac{nRT}{P}\,dP$$

$$\boxed{G - G° = nRT\,\ln(P/P°)}\ . \tag{3.1}$$

If we now consider the reaction

$$a\mathrm{A} + b\mathrm{B} \rightleftharpoons c\mathrm{C} + d\mathrm{D}$$

in which A, B, C, and D are all gases, we can apply the above equation to each component of the reaction. Collecting terms and rearranging this gives

$$\Delta G - \Delta G° = RT\,\ln \frac{(P_C/P_C°)^c (P_D/P_D°)^d}{(P_A/P_A°)^a (P_B/P_B°)^b} \tag{3.2}$$

where

$$\Delta G° = G°\ (\text{products}) - G°\ (\text{reactants}).$$

Suppose the reaction has come to *equilibrium*, then $\Delta G = 0$, and the values of P_A, etc., would be their *equilibrium values*. Hence

$$\boxed{-\Delta G° = RT\,\ln \frac{(P_C/P_C°)^c (P_D/P_D°)^d}{(P_A/P_A°)^a (P_B/P_B°)^b}}\ .$$

$\Delta G°$ is the change in free energy when 1 mole of the reactants is converted to 1 mole of products, each *in their standard states*. At a given temperature, this is a constant and so the R.H.S. of the equation must also be a constant. We rewrite the equation as

$$\boxed{-\Delta G° = RT\,\ln K_\mathrm{p}} \tag{3.3}$$

where K_p is known as the *equilibrium constant*, and is defined as

$$\boxed{K_\mathrm{p} = \frac{(P_C/P_C°)^c (P_D/P_D°)^d}{(P_A/P_A°)^a (P_B/P_B°)^b}}\ .$$

We have retained the terms in P° for each component in this equation (even though $P^\circ = 1$ atm). This emphasizes the point that the term for each component is the *ratio* of its pressure at equilibrium to that in the standard state. K_p is therefore dimensionless, although this point is often ignored by many workers.†

Applications to solutions

The equation linking ΔG° and K_p has been derived for reactions involving ideal gases. We can use a similar relationship for reactions in solution if we make the following modifications:

(a) Definitions of standard state

For *gases* the standard state is defined as the pure gas at 1 atmosphere pressure, while for *liquids* it is the pure liquid at 1 atmosphere pressure. For *solutes* (whether ionic or not) the standard state is usually defined as a 1 mol dm^{-3} solution, under a pressure of 1 atmosphere. We shall consider this in more detail in Chapter 5. For solids, the standard state is defined as the pure solid at 1 atmosphere pressure. Thus if a solid is present, it is always in its standard state, and hence terms involving the solid do not appear in the equilibrium constant expression.

(b) Definition of equilibrium constant

For solutions, the equilibrium constant is now defined in terms of concentrations rather than pressures. We shall denote the equilibrium constant by K.

These points are illustrated in the examples below.

Worked examples‡

(1) Consider the reaction

$$CaCO_3 \rightleftharpoons CaO + CO_2$$
$$\text{(s)} \qquad \text{(s)} \quad \text{(g)}$$

In this reaction $CaCO_3$ and CaO are both in their standard states, since they are solids (and will not appear in the expression for the equilibrium

---

† Throughout this book we shall treat equilibrium constants as being dimensionless, but shall include the standard state(s) referred to as a bracketed quantity, e.g. where other workers write $K = 5$ mol dm^{-3}, this would be written as $K = 5$ (mol dm^{-3}).

‡ In this book we have used natural logarithms. Students will recall that $\ln x \equiv \log_e x = 2{\cdot}303 \log_{10} x$.

constant). The standard state of CO_2 is 1 atmosphere. Hence the equilibrium constant becomes

$$K = \frac{(P_{CO_2})}{(P^\circ_{CO_2})}.$$

At 298 K the standard free energies of formation of $CaCO_3$, CaO, and CO_2 are $-1130\cdot3$, $-593\cdot6$, and $-394\cdot2$ kJ mol^{-1}, respectively. Thus ΔG° for the reaction is $+142\cdot5$ kJ mol^{-1}.

Using

$$\Delta G^\circ = -RT \ln K \qquad (3.3)$$

we have

$$\Delta G^\circ = -RT \ln(P_{CO_2}/P^\circ_{CO_2}),$$

i.e. $\ln(P_{CO_2}/P^\circ_{CO_2}) = -\dfrac{\Delta G^\circ}{RT}$

$$= -\frac{142\,500}{(8\cdot31)(298)}$$

$$= -57\cdot5$$

$$\therefore \quad (P_{CO_2}/P^\circ_{CO_2}) = 1\cdot07 \times 10^{-25}$$

$$\therefore \quad P_{CO_2} = 1\cdot07 \times 10^{-25} \text{ atm}, \quad \text{since } P^\circ_{CO_2} = 1 \text{ atm}.$$

(2) In the reaction

$$\text{citrate}^{3-} \rightarrow cis\text{-aconitate}^{3-} + H_2O$$

the standard free energies of formation at 298 K are $-1167\cdot1$, $-921\cdot7$, $-237\cdot0$ kJ mol^{-1} for citrate, cis-aconitate and H_2O. What is the equilibrium constant for the reaction, and what concentration of citrate will be in equilibrium with $0\cdot4$ mmol dm^{-3} cis-aconitate?

Solution

From the ΔG°_f values, the ΔG° for the reaction is $+8\cdot4$ kJ mol^{-1}. Using

$$-\Delta G^\circ = RT \ln K, \qquad (3.3)$$

we obtain

$$K = \underline{0\cdot034 \text{ (mol dm}^{-3})}$$

The true equilibrium constant for the reaction (K_{eq}) can be written

$$K_{eq} = \frac{([cis\text{-aconitate}]/[cis\text{-aconitate}]^\circ)([H_2O]/[H_2O]^\circ)}{([citrate]/[citrate]^\circ)},$$

where [ ] refers to concentration and [ ]$^\circ$ to the standard state concentration. In dilute solutions, however, the concentration of water is so large (55.5 mol dm^{-3}) that it can be considered to be effectively constant and the equilibrium constant for the reaction in aqueous solution (K) is obtained by dividing K_{eq} by the term $([H_2O]/[H_2O]^{\circ})$

$$K = \frac{([cis\text{-aconitate}]/[cis\text{-aconitate}]^{\circ})}{([\text{citrate}]/[\text{citrate}]^{\circ})}.$$

Since the standard states of the solutes citrate and *cis*-aconitate are 1 mol dm^{-3} solutions, this can be simplified to give:

$$K = \frac{[cis\text{-aconitate}]}{[\text{citrate}]}.$$

Now $K = 0.034$ (mol dm^{-3}) and $[cis\text{-aconitate}] = 0.4$ mmol dm^{-3}.

$$\therefore \quad [\text{citrate}] = 11.8 \text{ mmol dm}^{-3}.$$

Note. In general we do not write out the full expressions for the equilibrium constant (i.e. including the standard states), but use equations such as the simpler one for K above. This is illustrated again in the next example.

(3) A solution is 0.05 mol dm^{-3} in glyceraldehyde-3-P (G-3-P). After addition of triosephosphate isomerase the concentration of G-3-P was 0.002 mol dm^{-3} $(T = 298 \text{ K})$.

Calculate the ΔG° for the reaction catalysed by this enzyme, i.e. glyceraldehyde-3-P $\rightarrow$ dihydroxyacetone-P.

Solution

At equilibrium the [dihydroxyacetone-P] is 0.048 mol dm^{-3}. The equilibrium constant K for the reaction is given by

$$K = \frac{[\text{dihydroxyacetone-P}]}{[\text{glyceraldehyde-3-P}]}$$

$$= \frac{[0.048]}{[0.002]}$$

$$K = 24 \text{ (mol dm}^{-3}).$$

Substituting in:

$$\Delta G^{\circ} = -RT \ln K, \tag{3.3}$$

we have:

$$\Delta G^{\circ} = -8.31 \times 298 \times \ln 24$$

$$\underline{\Delta G^{\circ} = -7.87 \text{ kJ mol}^{-1}.}$$

(4) Under certain conditions, the intracellular levels of ADP and P_i were found to be 3 mmol dm^{-3} and 1 mmol dm^{-3} respectively. Assuming $\Delta G° = -30\cdot5$ kJ mol^{-1} for ATP hydrolysis, calculate the concentration of ATP which would be present at equilibrium ($T = 310$ K).

The actual level of ATP was found to be 10 mmol dm^{-3} under these conditions. What is the value of ΔG for the reaction

$$ATP \rightarrow ADP + P_i.$$

under these conditions?

Solution

Using the equation

$$\Delta G° = -RT \ln K \tag{3.3}$$

we have

$$K = 1\cdot3 \times 10^5 \text{ (mol dm}^{-3})$$

for the hydrolysis of ATP. The equilibrium constant for the reaction can be written as:

$$K = \frac{[ADP][P_i]}{[ATP]}.$$

Substituting $[ADP] = 3 \times 10^{-3}$ mol dm^{-3} and $[P_i] = 10^{-3}$ mol dm^{-3}

$$[ATP] = \frac{3 \times 10^{-6}}{1\cdot3 \times 10^5} \text{ mol dm}^{-3}$$

$$= 2\cdot3 \times 10^{-11} \text{ mol dm}^{-3}.$$

Since the actual level of ATP is 10 mmol dm^{-3} it is clear that the reaction is not at equilibrium. We can calculate the actual value of ΔG under intracellular conditions using the equation

$$\Delta G - \Delta G° = RT \ln \frac{[ADP][P_i]}{[ATP]}$$

which is the form of eqn (3.2) applicable to solutes,

$$= RT \ln \frac{3 \times 10^{-3} \times 1 \times 10^{-3}}{10 \times 10^{-3}}$$

$$= -20\cdot9 \text{ kJ mol}^{-1}.$$

$$\therefore \Delta G = -30\cdot5 - 20\cdot9 \text{ kJ mol}^{-1}$$

$$= \underline{-51\cdot4 \text{ kJ mol}^{-1}.}$$

If the level of ATP were $2\cdot3\times10^{-11}\,\text{mol dm}^{-3}$, the system would be at equilibrium and no work could be obtained from the reaction. Under the actual conditions we see that ΔG is even more negative than ΔG° and hence there is an even greater tendency for the reaction to occur, than if all the components were in their standard states (i.e. 1 mol dm^{-3}). This illustrates the point made previously that the further the system is from its equilibrium position, the greater the work (ΔG) that can be obtained. In the cell, the energy (work) made available by ATP hydrolysis is used in various ways (mechanical work, biosynthetic pathways, transport of metabolites against concentration gradients, etc.).

Distinction between ΔG and ΔG°

Before finishing this section we should perhaps just emphasize once again the distinction between ΔG and ΔG° for a reaction. ΔG° is the difference in free energy between products and reactants (each in their standard states). This does *not* refer to the actual reaction at equilibrium (except in the very unusual case where K, the equilibrium constant, equals one). ΔG, however, refers to the difference in free energy (products − reactants) at any other concentrations (or pressures) of the components. When $\Delta G = 0$, the reaction is at equilibrium, and the concentrations of the components are those which appear in the equilibrium constant expression, so that, for instance

$$-\Delta G^\circ = RT \ln \frac{[\text{ADP}]_{eq}[\text{P}_i]_{eq}}{[\text{ATP}]_{eq}}$$

(since $\Delta G = 0$).

When the reaction is not at equilibrium, the actual value of ΔG is calculated from

$$\Delta G - \Delta G^\circ = RT \ln \frac{[\text{ADP}][\text{P}_i]}{[\text{ATP}]}.$$

It may be helpful to have an idea of the magnitude of ΔG° associated with a given value of K. Table 3.1 illustrates this for a temperature of 310 K.

TABLE 3.1

ΔG°_{310} kJ mol^{-1}	+17·80	+11·86	+5·93	0	−5·93	−11·86	−17·80
K	10^{-3}	10^{-2}	10^{-1}	1	10^1	10^2	10^3

Note that K increases by a factor of 10 as ΔG° doubles. Thus even quite large errors in the determination of an equilibrium constant lead to relatively small errors in ΔG°.

The variation of the equilibrium constant with temperature

At equilibrium we have the relationship

$$-\Delta G^\circ = RT \ln K \tag{3.3}$$

$$\therefore \quad \ln K = -\frac{\Delta G^\circ}{RT}$$

$$\frac{d(\ln K)}{dT} = -\frac{1}{R} \cdot \frac{d(\Delta G^\circ / T)}{dT}. \tag{3.4}$$

We simplify the R.H.S. of eqn (3.4) as follows:
Using

$$\Delta G^\circ = \Delta H^\circ - T \, \Delta S^\circ$$

we have

$$\frac{\Delta G^\circ}{T} = \frac{\Delta H^\circ}{T} - \Delta S^\circ \tag{3.5}$$

so that by differentiation

$$\boxed{\frac{d(\Delta G^\circ / T)}{dT} = -\frac{\Delta H^\circ}{T^2}} \tag{3.6}$$

assuming ΔH° and ΔS° are independent of temperature over the range considered[†],
by substitution in eqn (3.4)

$$\boxed{\frac{d(\ln K)}{dT} = \frac{\Delta H^\circ}{RT^2}}. \tag{3.7}$$

This eqn (3.7) is often referred to as the *Van't Hoff Isochore*. It is more useful in the form obtained by integrating between temperatures T_1 and T_2, viz:

$$\int_{K_1}^{K_2} d(\ln K) = \int_{T_1}^{T_2} \frac{\Delta H^\circ}{RT^2} \, dT \tag{3.8}$$

---

† This formula can be derived more generally *without* assuming that ΔS° is constant. The variation of enthalpy and entropy with temperature and pressure is discussed in Appendix 1.

$$\therefore \quad \ln \frac{K_2}{K_1} = -\frac{\Delta H^\circ}{R}\left(\frac{1}{T_2} - \frac{1}{T_1}\right).$$

(3.9)†

Thus a plot of $\ln K$ against $1/T$ gives a straight line of slope $-\Delta H^\circ/R$. This equation can be used in two main ways:

(1) If we know the values of K at various temperatures, ΔH° can be obtained from the slope of the graph of $\ln K$ vs. $1/T$;

(2) If we know K at one temperature and the value of ΔH°, we can calculate K at another temperature.

The assumptions that ΔH° and ΔS° are both independent of temperature are generally satisfactory, if the temperature range considered is fairly small. For more accurate work variations in these quantities with temperature can be taken into account (see Appendix 1) in the integrated expression. Provided the plot of $\ln K$ vs. $1/T$ is linear we can be confident about the validity of these assumptions.

We now consider some examples of the use of this equation.

Worked examples

(1) The equilibrium constant for a reaction doubles on raising the temperature from 298 K to 308 K. What is the value of ΔH° for the reaction?

Solution

Using eqn (3.9), i.e.

$$\ln \frac{K_2}{K_1} = -\frac{\Delta H^\circ}{R}\left(\frac{1}{T_2} - \frac{1}{T_1}\right).$$

---

† A simpler way of obtaining this equation is to combine eqn (3.3) with eqn (3.5).

$$\ln K = -\frac{\Delta G^\circ}{RT}$$

$$\ln K = -\frac{\Delta H^\circ}{RT} + \frac{\Delta S^\circ}{R}$$

Assuming that ΔH° and ΔS° are constant between two temperatures, T_1 and T_2, we can then write:

$$\ln \frac{K_2}{K_1} = -\frac{\Delta H^\circ}{R}\left(\frac{1}{T_2} - \frac{1}{T_1}\right)$$

which is eqn (3.9).

Now $K_2/K_1 = 2$, $T_2 = 308$ K, and $T_1 = 298$ K.

$$\therefore \quad \ln 2 = -\frac{\Delta H^\circ}{R}\left(\frac{1}{308} - \frac{1}{298}\right)$$

$$\therefore \quad \Delta H^\circ = 52{\cdot}9 \text{ kJ mol}^{-1}.$$

This gives us a useful 'rule of thumb'—if an equilibrium constant is doubled for a 10 K temperature rise, the ΔH° is about 53 kJ mol^{-1}. If the equilibrium constant is not so sensitive to temperature, ΔH° is less than this value.

We should also note that if K *increases with increasing temperature* ΔH° is *positive*. This is easily seen by considering the eqn (3.7), i.e.

$$\frac{\mathrm{d}\ln K}{\mathrm{d}T} = \frac{\Delta H^\circ}{RT^2}.$$

(2) The following data were obtained for the temperature variation of the equilibrium constant of an inhibitor binding to the enzyme carbonic anhydrase. Determine the value of ΔG°, ΔH°, and ΔS° for this process at 298 K.

T/K	289	294·2	298	304·9	310·5
$K(\times 10^{-7})(\text{mol dm}^{-3})$	7·25	5·25	4·17	2·66	2·0

Solution

We can express the data in the appropriate form:

T/K	289	294·2	298	304·9	310·5
$1/T \times 10^3$	3·460	3·400	3·356	3·280	3·221
$K \times 10^{-7}$ (mol dm^{-3})	7·25	5·25	4·17	2·66	2·0
$\ln K$	18·10	17·78	17·55	17·10	16·81

From the slope of $\ln K$ vs. $1/T$ we can calculate that $\Delta H^\circ = -45{\cdot}1$ kJ mol^{-1}. Now, at 298 K, $K = 4{\cdot}17 \times 10^7$ so that using the equation

$$-\Delta G^\circ = RT \ln K, \tag{3.3}$$

we can calculate that

$$\Delta G^\circ_{298} = -43{\cdot}45 \text{ kJ mol}^{-1}.$$

Now

$$\Delta G^\circ = \Delta H^\circ - T\Delta S^\circ,$$

$$\therefore \quad \Delta S^\circ = \frac{\Delta H^\circ - \Delta G^\circ}{T},$$

$$\Delta S^\circ = \frac{(-45 \cdot 1 + 43 \cdot 45)}{298} \, 1000,$$

i.e.

$$\Delta S^\circ = -5 \cdot 53 \text{ J K}^{-1} \text{ mol}^{-1}.$$

(3) At 298 K the value of the equilibrium constant for the binding of phosphate to aldolase is 540 (mol dm^{-3}). Direct measurements show that the enthalpy change is $-87 \cdot 8$ kJ mol^{-1}. What is the value of the equilibrium constant at 310 K if ΔH° is assumed independent of temperature?

Solution

Using eqn (3.9)

$$\ln \left(\frac{K_2}{K_1} \right) = -\frac{\Delta H^\circ}{R} \left(\frac{1}{T_2} - \frac{1}{T_1} \right),$$

$$K_1 = 540, \, \Delta H^\circ = -87\,800 \text{ J mol}^{-1}, \, T_2 = 310 \text{ K}, \, T_1 = 298 \text{ K},$$

$$\therefore \quad \ln \left(\frac{K_2}{540} \right) = -\frac{(87\,800)}{8 \cdot 31} \left(\frac{1}{310} - \frac{1}{298} \right),$$

$$= -1 \cdot 372,$$

$$\therefore \quad \frac{K_2}{540} = 0 \cdot 253,$$

$$K_2 = 137,$$

i.e. the equilibrium constant at 310 K is $\underline{137}$ (mol dm^{-3}).

This is in accord with our expectations, since ΔH° is negative and so K should decrease with increasing temperature. Also since ΔH° is numerically greater than 53 kJ mol^{-1}, the equilibrium constant should change by a factor greater than 2 for this change in temperature, as indeed is the case.

Measurements of the thermodynamic functions of reactions

We are in a position to indicate some of the methods for measuring ΔG°, ΔH°, and ΔS° for a reaction.

Measurement of ΔG°

If the equilibrium constant for a reaction can be measured, ΔG° may be calculated directly from the equation:

$$-\Delta G^\circ = RT \ln K.$$

In some cases the equilibrium lies so far over to one side that direct measurement of K is difficult. ΔG° can then be obtained either by knowing

$\Delta H°$ and $\Delta S°$ (as outlined below) *or* by making measurements on associated or coupled reactions for which the $\Delta G°$s are known. (See Chapter 2.) Of course, if the values of the standard free energies of formation of the components of the reaction are available in Tables or can be measured, then $\Delta G°$ for the reaction is readily calculated.

Measurement of $\Delta H°$

The value for the enthalpy change of a reaction can be determined by measuring the equilibrium constant at various temperatures and then using the equation:

$$\ln \left(\frac{K_2}{K_1}\right) = -\frac{\Delta H°}{R}\left(\frac{1}{T_2} - \frac{1}{T_1}\right). \tag{3.9}$$

As described above a plot of $\ln K$ vs. $1/T$ gives a straight line of slope $-\Delta H°/R$.

Alternatively, $\Delta H°$ may be obtained by direct calorimetric measurements at constant pressure. If measurements are made at constant volume (as for instance in a bomb calorimeter) a correction has to be made, as outlined in Chapter 1. This direct determination of $\Delta H°$ is becoming more widespread because of the availability of extremely accurate microcalorimeters.

Again if the standard enthalpies of formation of the components of the reaction are available or can be measured (e.g. using combustion data), the $\Delta H°$ for the reaction can be calculated.

Measurement of $\Delta S°$

$\Delta S°$ is usually obtained by knowing $\Delta G°$ and $\Delta H°$ at a certain temperature and then substituting in the equation

$$\Delta G° = \Delta H° - T\,\Delta S°.$$

However $\Delta S°$ can also be calculated if the absolute standard entropies of the components at the temperature in question are known. These are derived using the Third Law of Thermodynamics as outlined in Appendix 1.

A knowledge of $\Delta G°$, $\Delta H°$, and $\Delta S°$ for a reaction enables us to predict not only the value of the equilibrium constant at a given temperature, but also the variation of this constant with temperature. This is the principal application of thermodynamics with which we are concerned.

PROBLEMS

1. What is the difference between ΔG and $\Delta G°$ for a reaction? Calculate $\Delta G°$ for the following processes:
   (a) $H_2O\ (1, 373\cdot15\ \text{K}) \rightleftharpoons H_2O\ (g, 373\cdot15\ \text{K})$.
   (b) $H_2O\ (1, 310\ \text{K}) \rightleftharpoons H_2O\ (g, 310\ \text{K})$.
   The vapour pressure of H_2O at 310 K is $0\cdot062$ atm.

2. $\Delta G^{\circ\prime}$ for the reaction

$$\text{glycerol} + P_i \rightleftharpoons \text{L-glycerol-1-phosphate} + H_2O$$

is $11 \cdot 1$ kJ mol^{-1} at 298 K. What concentration of L-glycerol-1-phosphate will be present at equilibrium if we start with 1 mol dm^{-3} glycerol and $0 \cdot 5$ mol dm^{-3} phosphate?

3. The conversion of glucose-6-P to fructose-6-P is catalysed by the enzyme phosphoglucose isomerase. For this reaction

$$\text{G-6-P} \rightleftharpoons \text{F-6-P}; \qquad \Delta G^{\circ\prime} = 2 \cdot 1 \text{ kJ mol}^{-1}.$$

If we start with $0 \cdot 1$ mol dm^{-3} G-6-P what is the final composition of the solution? ($T = 298$ K.)

If phosphoglucomutase, which catalyses the following reaction, is now added,

$$\text{G-1-P} \rightleftharpoons \text{G-6-P}; \qquad \Delta G^{\circ\prime} = -7 \cdot 27 \text{ kJ mol}^{-1}$$

what is the new composition of the mixture at equilibrium?

4. The formation of G-6-P from glucose and P_i is unfavourable.

$$\text{glucose} + P_i \rightleftharpoons \text{G-6-P} + H_2O; \qquad \Delta G^{\circ\prime} = +17 \cdot 1 \text{ kJ mol}^{-1}$$

whereas the hydrolysis of phosphoenolpyruvate (PEP) is highly favourable

$$H_2O + PEP \rightleftharpoons \text{pyruvate} + P_i; \qquad \Delta G^{\circ\prime} = -55 \cdot 2 \text{ kJ mol}^{-1}.$$

Show how these reactions can be coupled to the synthesis or hydrolysis of ATP, given that

$$H_2O + ATP \rightleftharpoons ADP + P_i; \qquad \Delta G^{\circ\prime} = -30 \cdot 5 \text{ kJ mol}^{-1}.$$

Calculate the equilibrium constants for these coupled reactions. ($T = 310$ K.)

5. Glycogen phosphorylase catalyses the reaction:

$$(\text{glycogen})_n + P_i \rightleftharpoons (\text{glycogen})_{n-1} + \text{G-1-P}.$$

Assuming that the concentrations of P_i and G-1-P are equal, does the equilibrium lie in favour of glycogen synthesis or degradation? ($T = 310$ K, $\Delta G^{\circ\prime} = +3 \cdot 05$ kJ mol^{-1}.)

In muscle, the concentrations of P_i and G-1-P are found to be 10 mmol dm^{-3} and 30 μmol dm^{-3}. On which side does the equilibrium lie in muscle?

6. In a study of the reaction catalysed by phosphofructokinase

$$\text{F-6-P} + ATP \rightleftharpoons FDP + ADP$$

the following data were obtained for metabolite levels in perfused rat heart at 308 K.

F-6-P	60 μmol dm^{-3}
ATP	$5 \cdot 3$ mmol dm^{-3}
FDP	9 μmol dm^{-3}
ADP	$1 \cdot 1$ mmol dm^{-3}

$\Delta G^{\circ\prime}$ for the phosphofructokinase reaction is $-17 \cdot 7$ kJ mol^{-1}. Is the reaction at equilibrium in the perfused heart? If not, what is the value of $\Delta G'$?

In the same tissue the concentration of AMP was found to be 95 μmol dm^{-3}. Is the reaction catalysed by adenylate kinase

$$2ADP \rightleftharpoons AMP + ATP$$

at equilibrium? ($\Delta G^{\circ\prime} = +2\cdot1$ kJ mol^{-1}.)
If not, what is the value of $\Delta G'$?
Comment briefly on your results.

7. Calculate $\Delta G^{\circ\prime}$ for the hydrolysis of L-leucylglycine by peptidase at 298 K. The standard Gibbs free energies of formation in kJ mol^{-1} of the compounds in aqueous solution at 298 K are:

L-leucylglycine	-464	water	-237
glycine	-373	L-leucine	-341

In vivo, new proteins are synthesized by activating amino acids by attachment to the appropriate tRNA. This is accomplished at the expense of ATP hydrolysis. Using the above value of $\Delta G^{\circ\prime}$ for peptide bond hydrolysis, and the following information:

$$(\text{amino acid})_1 - tRNA + ATP \rightleftharpoons (\text{amino acyl})_1 - tRNA + AMP + 2P_i;$$

$$\Delta G^{\circ\prime} = -29\cdot3 \text{ kJ mol}^{-1}.$$

$$H_2O + ATP \rightleftharpoons AMP + 2P_i; \qquad \Delta G^{\circ\prime} = -61\cdot0 \text{ kJ mol}^{-1}.$$

Calculate $\Delta G^{\circ\prime}$ for the reaction step involved in protein biosynthesis, i.e.

$$(\text{amino acyl})_1 - tRNA + (\text{amino acid})_2 \rightleftharpoons (\text{amino acid})_1 - (\text{amino acid})_2 + tRNA.$$

What is the equilibrium constant for this reaction?

8. For the oxygenation of haemoglobin (Hb)

$$Hb\,(aq) + O_2(g) \rightleftharpoons HbO_2\,(aq)$$

The equilibrium constant is $85\cdot5$ at 292 K.
   Calculate ΔG° for this process.
   At 292 K, in equilibrium with air, the partial pressure of oxygen is 0.2 atm and the solubility of oxygen in water is $0\cdot23$ mmol dm^{-3}. From these data calculate ΔG° for the reaction

$$O_2(g) \rightleftharpoons O_2(aq)$$

and hence ΔG° for

$$Hb(aq) + O_2(aq) \rightleftharpoons HbO_2(aq)$$

What assumptions are made in these calculations?

9. The equilibrium constant for the hydrolysis of ATP at 310 K is $1\cdot3 \times 10^5$ (mol dm^{-3})

$$H_2O + ATP \rightleftharpoons ADP + P_i; \qquad K = 1\cdot3 \times 10^5 \text{ (mol dm}^{-3}).$$

If $\Delta H^{\circ\prime} = -20\cdot0$ kJ mol^{-1}, calculate K at 298 K and 273$\cdot$15 K.
In the mitochondria, ATP synthesis is thought to be achieved by removal of H_2O from the active site of the ATPase enzyme. If the P_i concentration is

$10 \, \text{mmol dm}^{-3}$, at what concentration of water would the ATP and ADP concentrations be equal? ($T = 310$ K.)

10. The enzyme phosphorylase b can be activated by the addition of AMP. The dissociation constant (K_d) for the reaction:

$$\text{phosphorylase b.AMP} \rightleftharpoons \text{phosphorylase b} + \text{AMP}$$

varied with temperature as follows:

Temperature/K	285·5	289	300	312·5
$K_d(\times 10^5) \, (\text{mol dm}^{-3})$	2·75	3·1	4·2	5·9

Calculate $\Delta H°$, $\Delta S°$, and $\Delta G°$ for this reaction at 303 K.
State any assumptions you have to make.

11. In the biosynthesis of proteins it is important that the correct tRNA binds to its corresponding amino acid tRNA ligase enzyme. In the case of isoleucine tRNA ligase from *E. coli*, the binding of isoleucine tRNA under certain conditions is characterized by values of $\Delta H° = 0 \, \text{kJ mol}^{-1}$ and $\Delta S° = +142 \, \text{J K}^{-1} \text{mol}^{-1}$. The binding of valine tRNA under these conditions is characterized by values of $\Delta H° = +33·4 \, \text{kJ mol}^{-1}$ and $\Delta S° = +225·7 \, \text{J K}^{-1} \text{mol}^{-1}$. By what factor is binding of the correct tRNA favoured over incorrect tRNA at 293 K and at 313 K?

12. At 293 K, the $\Delta H°$ for the association of enzyme subunits was found to be $+16·8 \, \text{kJ mol}^{-1}$. The $\Delta S°$ was $+66·9 \, \text{J K}^{-1} \text{mol}^{-1}$. Would association be favourable at this temperature?

$$2 \, (\text{isolated subunits}) \rightleftharpoons \text{dimer.}$$

What would be the effect of lowering the temperature? Does this tell us anything about the phenomenon of 'cold lability' of several associated (i.e. multisubunit) enzymes?

4. Binding of ligands to macromolecules

Introduction

IT is convenient to discuss the analysis of ligand binding data in a separate section, as several additional points are brought out although no new fundamental principles are involved.

Let us consider, initially, the case where one mole of A combines with one mole of P

$$P + A \rightleftharpoons PA.$$

Then the equilibrium constant for this association reaction is

$$K = \frac{[PA]}{[P][A]} \quad \text{where [ ] represent concentrations.}$$

Biochemists usually refer to a 'dissociation constant' (often given the symbol K_d) which is merely the reciprocal of the above association constant

$$\text{i.e.} \quad K_d = \frac{[P][A]}{[PA]}. \tag{4.1}$$

The binding equation—treatment of binding data

Let us define the number of moles of A bound per mole of P under a given set of conditions as r.† Then

$$r = \frac{\text{concentration of A bound to P}}{\text{total concentration of all forms of P}} \tag{4.2}$$

$$= \frac{[PA]}{[P] + [PA]}. \tag{4.3}$$

Now, from eqn (4.1)

$$[PA] = \frac{[P][A]}{K_d}.$$

So that

$$r = \frac{([P][A])/K_d}{[P] + ([P][A])/K_d}, \tag{4.4}$$

i.e.

$$\boxed{r = \frac{[A]}{K_d + [A]}}. \tag{4.5}$$

---

† In this case, $r \le 1$ and is the 'fractional saturation of the sites'. This is *not* the case, of course when we consider multiple binding sites (p. 37).

Note that when $r = 0.5$ (i.e. P is half saturated with A) then $[A] = K_d$.
The fundamental eqn (4.5) can be rearranged into several forms suitable for graphical treatment. Two of the most common are dealt with below:
(1) The equation is rearranged to give

$$\frac{1}{r} = 1 + \frac{K_d}{[A]}.$$

(4.6)

So that a graph of $1/r$ against $1/[A]$ gives a straight line of slope K_d.
(2) Alternatively the equation can be rearranged to give

$$\frac{r}{[A]} = \frac{1}{K_d} - \frac{r}{K_d}.$$

(4.7)

A graph of $r/[A]$ against r gives a straight line whose slope is $-1/K_d$.
Thus if, from experimental data the amount of A bound to P can be calculated (at given total concentrations of P and A), we can evaluate the dissociation constant. It is important to note that in these equations [A] refers to the concentration of A which is free in solution, i.e. not bound to P.

Worked example

Mg^{2+} and ADP form a 1:1 complex. In an experiment, the concentration of ADP was kept constant at 80 μmol dm^{-3} and the concentration of Mg^{2+} varied. The following results were obtained.

Total Mg^{2+} (μmol dm^{-3})	20	50	100	150	200	400
Mg^{2+} bound to ADP (μmol dm^{-3})	11·6	26·0	42·7	52·8	59·0	69·5

Determine the dissociation constant for MgADP under these conditions.

Solution

At each value of the total Mg^{2+} concentration, the free Mg^{2+} concentration ([A] in the equations) can be evaluated simply by difference. The value of r is found by dividing the concentration of bound Mg^{2+} by the ADP concentration (i.e. 80 μmol dm^{-3}). We can convert the data into the correct form for graphical treatment.

Total Mg^{2+} (μmol dm^{-3})	20	50	100	150	200	400
Bound Mg^{2+} (μmol dm^{-3})	11·6	26·0	42·7	52·8	59·0	69·5
Free Mg^{2+} (μmol dm^{-3})	8·4	24·0	57·3	97·2	141·0	330·5
r	0·145	0·325	0·534	0·660	0·738	0·869
$\dfrac{1}{r}$	6·90	3·08	1·874	1·515	1·356	1·151
$\dfrac{1}{[Mg^{2+}]_{free}}$ (μmol dm^{-3})$^{-1}$	0·1190	0·0417	0·0175	0·0103	0·0071	0·0030
$\dfrac{r}{[Mg^{2+}]_{free}}$ (μmol dm^{-3})$^{-1}$	0·0173	0·0135	0·0093	0·0068	0·0052	0·0026

The appropriate plots ($1/r$ vs. $1/[Mg^{2+}]_{free}$ and $r/[Mg^{2+}]_{free}$ vs. r) are shown in Figs. 4.1 and 4.2 respectively.

Of course we would not normally do both, but this is done here for the sake of completeness.

From both plots we obtain the result that $K_d = 50 \ \mu mol \ dm^{-3}$ or $50 \times 10^{-6} \ (mol \ dm^{-3})$.[†] It is noticeable that in the 'double reciprocal plot' (Fig. 4.1) the experimental points are much more unevenly spaced than in the alternative plot (Fig. 4.2). This has led many workers to prefer the type of plot shown in Fig. 4.2 for the analysis of binding data, since it is rather easier in this case to draw the best straight line through the experimental points. In any experiment it is important to make a proper analysis of the distribution of errors in the method of plotting the data. This is also true in the analogous plots which are used in the analysis of enzyme kinetic data (see Chapter 10) and is discussed in the books by Cornish-Bowden mentioned in the reading list.

It is possible to simplify the experiment considerably if one component is present in a considerable excess over the other. For instance, suppose that P is present at a concentration of $1 \ \mu mol \ dm^{-3}$ and [A] is varied from $50 \ \mu mol \ dm^{-3}$ up to $500 \ \mu mol \ dm^{-3}$. Then, throughout the titration very little of the total A is actually bound to P and it is a very good approximation to write $[A]_{free} = [A]_{total}$. The equations would then become

$$\frac{1}{r} = 1 + \frac{K_d}{[A]_{total}} \quad and \quad \frac{r}{[A]_{total}} = \frac{1}{K_d} - \frac{r}{K_d}.$$

We often use this simplification in enzyme kinetic work. The substrate (S) of the enzyme is almost always greatly in excess over the enzyme concentration. (i.e. $[S]_{free} = [S]_{total}$.) In this type of work, we use the velocity of the enzyme catalysed reaction (v) to give a measure of r (the amount of S bound to E) in the equations, since only the ES complex shows enzyme activity. We shall see in Chapter 10 that we do in fact plot $1/v$ vs. $1/[S]_{total}$ or $v/[S]_{total}$ vs. v to obtain the *Michaelis constant* which characterizes the interaction of the enzyme with its substrate.[‡]

The simplification of the algebra which is achieved by setting $[S]_{free}$ equal to $[S]_{total}$ is illustrated in the following example.

Worked example

Consider the equilibrium $E + S \rightleftharpoons ES$, and let K_s be the dissociation constant of the ES complex.

[†] Strictly speaking, K_d is dimensionless, as is pointed out earlier in Chapter 3. However, biochemists invariably quote units, i.e. $K_d = 50 \ \mu mol \ dm^{-3}$. Referred to a $1 \ mol \ dm^{-3}$ standard state, we could say $K_d = 50 \times 10^{-6}$. We shall adopt the convention of writing dissociation constants as for example, $50 \times 10^{-6} \ (\mu mol \ dm^{-3})$ where the bracketed quantity refers to the standard state of a $1 \ mol \ dm^{-3}$ solution.

[‡] *Cautionary note*: the Michaelis constant (K_m) is not generally a true dissociation constant (see Chapter 10).

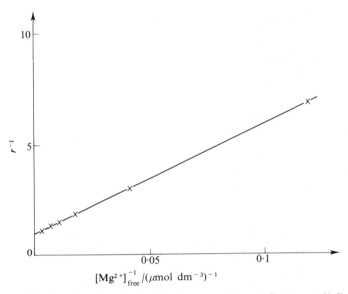

FIG. 4.1. Plot of binding data in 'Worked example' according to eqn (4.6).

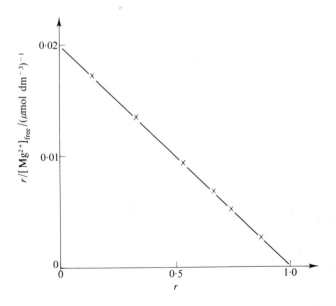

FIG. 4.2. Plot of binding data in 'Worked example' according to eqn (4.7).

What is the fraction of enzyme present as ES if

(a) $K_s = 50 \times 10^{-6}$ (mol dm^{-3}), $[E]_{total} = 100$ μmol dm^{-3}, $[S]_{total}$
$$= 150 \ \mu mol \ dm^{-3},$$
(b) $K_s = 50 \times 10^{-6}$ (mol dm^{-3}), $[E]_{total} = 1$ μmol dm^{-3}, $[S]_{total}$
$$= 150 \ \mu mol \ dm^{-3}$$

Solution

In the case (a) $[E]_{total}$ is comparable with $[S]_{total}$. We must therefore solve a quadratic equation to calculate $[ES]$.

Let x μmol dm^{-3} be the concentration of ES

then $(100 - x)$ μmol dm^{-3} is the concentration of free E

and $(150 - x)$ μmol dm^{-3} is the concentration of free S,

$$\therefore \quad K_s = 50 \times 10^{-6} = \frac{(100 - x)(150 - x)}{x} . \ 10^{-6}$$

$$50x = (100 - x)(150 - x)$$

$$\therefore \quad x^2 - 300x + 15\ 000 = 0,$$

$$\therefore \quad x = \underline{63 \cdot 4.}$$

The total concentration of enzyme is 100 μmol dm^{-3} so in this case $63 \cdot 4$ per cent of the enzyme is present as ES.

In case (b)$[E]_{total} \ll [S]_{total}$ and thus $[S]_{free}$ can be set equal to $[S]_{total}$

$$\therefore \quad K_s = 50 \times 10^{-6} = \frac{[E][S]_{total}}{[ES]}$$

$$= \frac{[E]}{[ES]} . \ 150 \times 10^{-6}.$$

Thus $[E]/[ES] = \frac{1}{3}$, and hence $[ES]$ is 75 per cent of the total enzyme. An extension of this worked example is set as Problem 2.

Multiple binding site equilibria

So far we have only considered the case where one mole of a ligand (A) is bound by a macromolecule (P). However, there are many examples known where several moles of A can be bound by one mole of P. We now wish to examine these cases and to discuss equations which will enable us to determine the number of binding sites in these cases.

If we have a situation where the binding of ligand to macromolecule is very tight we can determine the number of binding sites by a direct titration. Aliquots of ligand are added to a known concentration of the macromolecule. When the amount of ligand added exceeds the concentration of binding sites available, the added ligand remains free in solution.

A typical application of this method is illustrated in Problem 3.

In cases where the binding is not sufficiently tight we must consider the appropriate equilibrium equations.

For a single binding site, we derived the equation $r = [A]/(K_d + [A])$, where r is the number of moles of A bound per mole of P, $[A]$ is the concentration of free A and K_d is the dissociation constant.

For the case where one mole of P can bind up to n moles of A this equation becomes modified:

$$r = \frac{n[A]}{K_d + [A]}.$$ (4.8)

where the n sites are assumed to be equal and independent (i.e. the free energy of binding is the same for each site). K_d is now an average dissociation constant.

The derivation of this equation is rather complex and is dealt with in Appendix 2.

As before, we can rearrange this equation in two main ways

(1) By taking the reciprocal of each side of the equation we derive

$$\frac{1}{r} = \frac{1}{n} + \frac{K_d}{n[A]}.$$ (4.9)

Thus a plot of $1/r$ against $1/[A]$ gives a straight line of slope K_d/n and an intercept on the y axis of $1/n$. This is often referred to as the Hughes–Klotz equation

(2) The alternative rearrangement is

$$\frac{r}{[A]} = \frac{n}{K_d} - \frac{r}{K_d},$$ (4.10)

so that a plot of $r/[A]$ against r gives a straight line of slope $-1/K_d$ and an intercept on the x-axis of n. This is often termed the Scatchard equation.

These two types of plots are illustrated in the worked example below.

Worked example

In an experiment the concentration of an enzyme is kept constant at 11 μmol dm^{-3}, and the concentration of inhibitor [I] varied. The following results were obtained.

$[I]_{total}(\mu mol\ dm^{-3})$	5·2	10·4	15·6	20·8	31·2	41·6	62·4
$[I]_{free}(\mu mol\ dm^{-3})$	2·3	4·8	7·95	11·3	18·9	27·4	45·8

Determine the dissociation constant for the enzyme-inhibitor complex and the number of inhibitor binding sites on the enzyme.

Solution

At each value of $[I]_{total}$ we can evaluate $[I]_{bound}$ by subtraction; r is obtained by dividing $[I]_{bound}$ by the concentration of enzyme (i.e. 11 μmol dm^{-3}). The following table can be constructed:

$[I]_{total}\ (\mu mol\ dm^{-3})$	5·2	10·4	15·6	20·8	31·2	41·6	62·4
$[I]_{free}\ (\mu mol\ dm^{-3})$	2·3	4·8	7·95	11·3	18·9	27·4	45·8
$[I]_{bound}\ (\mu mol\ dm^{-3})$	2·9	5·6	7·65	9·5	12·3	14·2	16·6
r		0·264	0·510	0·695	0·864	1·118	1·291	1·510
$\dfrac{1}{r}$		3·793	1·964	1·438	1·158	0·894	0·775	0·663
$\dfrac{r}{[I]_{free}}\ (\mu mol\ dm^{-3})^{-1}$	0·115	0·106	0·087	0·076	0·059	0·047	0·033
$\dfrac{1}{[I]_{free}}\ (\mu mol\ dm^{-3})^{-1}$	0·435	0·208	0·126	0·088	0·053	0·036	0·022

The two binding plots are shown in Figs. 4.3 and 4.4 respectively. From the 'double reciprocal plot' we find that the intercept on the y axis is 0·5, so that $n = 2$. The slope of the line is 7·6 so that $K_d = 15·2 \times 10^{-6}$ (mol dm^{-3}).

From the 'Scatchard' plot (Fig. 4.4), again we find that $n = 2$ and the value of K_d is $15·2 \times 10^{-6}$ (mol dm^{-3}). It is also clear that the sites are equivalent and independent, since, otherwise a curved plot would be expected.

As in the previous example (Mg^{2+} and ADP) we find that the Scatchard plot has a more even spacing of the experimental points, than does the 'double reciprocal'. (However, this need not always be the case.)

It is important to note that in order to determine the number of binding sites n accurately it is essential to cover as wide a range of the total saturation curve as possible. Roughly, the required range is the region

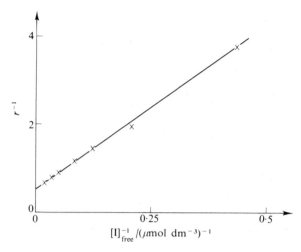

FIG. 4.3. Plot of binding data in 'Worked example' according to eqn (4.9).

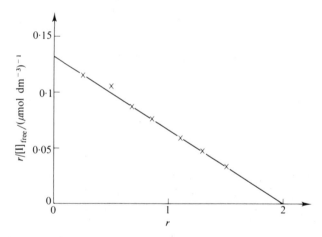

FIG 4.4. Plot of binding data in 'Worked example' according to eqn (4.10).

between 25 per cent saturation and 75 per cent saturation of the sites on the macromolecule for the ligand.

Non-equivalent ligand sites on a macromolecule

In the example discussed above, it is clear that there are two equivalent sites on the enzyme for the inhibitor. However, it often happens that the binding plots do not yield straight lines, and this reflects the fact that the ligand sites are not equivalent. An early example of this situation was found in the case of oxygen binding to haemoglobin (Hb) where the saturation curve is sigmoid in shape and should be compared with the curve for myoglobin (Mb) which is described as hyperbolic (i.e. follows a rectangular hyperbola†) (Fig. 4.5). Many other examples of this type of effect have since been found, both for binding proteins and for enzymes.

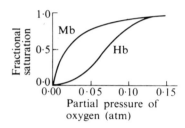

FIG. 4.5. Oxygen saturation curves for haemoglobin (Hb) and myoglobin (Mb).

Myoglobin consists of a single polypeptide chain, whereas haemoglobin consists of four subunits held together by non-covalent bonds, and it has become clear that the sigmoidal shapes of the saturation curves for haemoglobin and for other proteins are a consequence of a multisubunit structure. Haemoglobin would be said to show 'positive co-operativity' in binding of oxygen (i.e. binding of the first oxygen molecule enhances the binding of the subsequent ones). The 'advantage' of a sigmoid saturation curve is that over certain concentration ranges the fractional saturation of a molecule (or the rate of an enzyme catalysed reaction) can more readily respond to changes in ligand (or substrate) concentration, leading to greater possibilities of control of metabolic activity in the cell. Cases are also known of 'negative co-operativity' (in which the binding of the first ligand molecule discourages binding of subsequent ones). In this case, the fractional saturation or reaction rate would be less sensitive to changes in ligand or substrate concentration over certain concentration ranges.

The type of binding curve can be recognized from the saturation curve, the double reciprocal plot or the Scatchard plot. Typical curves are shown in Fig. 4.6 (these are not of quantitative significance, but only designed to show the shape of the curves).‡

---

† A rectangular hyperbola is characterized by an equation of the form of eqn (4.5).
‡ It should be emphasized that deviations from hyperbolic binding curves may be more easily recognized by the use of secondary plots (e.g. Fig. 4.6(b) and (c)).

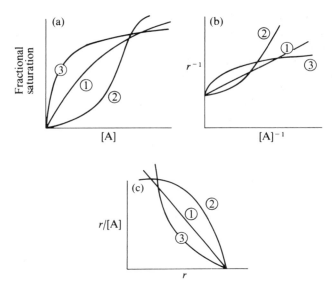

FIG. 4.6. Typical plots for binding data (a) saturation curve, (b) double reciprocal plot, (c) Scatchard plot. r represents the moles of A bound per mole of macromolecule and [A] represents the concentration of free ligand. Curves ①, ②, and ③ are typical for hyperbolic binding, positive co-operativity, and negative co-operativity respectively. Note in Fig. 4.6(a) it is difficult to distinguish between curves 1 and 3 hence the use of other plots.

Two principal theories to account for the sigmoid shape of saturation curves (e.g. for haemoglobin) have been proposed. Both stress the importance of changes in protein shape (or conformation) which can be associated with the binding of ligands to proteins. The 'concerted' theory of Monod, Wyman, and Changeux is based on the proposition that when a ligand binds to one subunit in a multisubunit protein all the subunits change their shape in a concerted fashion. In the other, 'sequential', model of Koshland, Nemethy, and Filmer, the conformational changes are assumed to be sequential, i.e. the ligand changes the conformation of only the subunit to which it binds. The conformational change will, of course, alter the interaction(s) between this subunit and its neighbouring subunit(s) making the binding of ligand to the latter stronger or weaker.

Although a distinction between the theories is not always easy (or valuable) to make, it is clear that the 'concerted' model cannot apparently account for the phenomenon of negative co-operativity (such as is apparently observed, for instance, in the binding of NAD^+ to rabbit muscle glyceraldehyde-3-phosphate dehydrogenase). However, alternative explanations to negative co-operativity can be involved (such as the binding sites being non-equivalent in the absence of ligands). The reader is referred to more detailed treatments of this subject (such as that found in the book by Newsholme and Start referred to in the reading list).

Experimental methods for obtaining binding data

This aspect of the subject will not be discussed in detail here either. There are two principal types of technique employed in these studies.

(1) The macromolecule with its bound ligand is separated from the free ligand (usually on the basis of size).† The concentration of free ligand can be readily evaluated, and hence the concentration of bound ligand is determined by subtraction of this number from the known total ligand concentration.

(2) We can use an indirect method to monitor binding—e.g. by looking for a spectroscopic change in either the macromolecule or the ligand which distinguishes the free macromolecule (or ligand) from the complex between the two. Under any given set of conditions we can then determine the contributions of the free and bound forms.

Using either of these types of techniques we were able to evaluate the parameters r and $[A]$ in eqn (4.8). The data can then be plotted either in the 'double reciprocal' or in the 'Scatchard' form to determine the parameters n and K.

PROBLEMS

1. The binding of a substrate (ADP) to the enzyme pyruvate kinase was studied by measuring the quenching of the protein fluorescence. The following results were obtained.

Total concentration of ADP (mmol dm^{-3})	0·29	0·36	0·42	0·53	0·84	1·18	1·89
mol ADP bound/mol protein	1·25	1·35	1·62	1·82	2·26	2·62	3·03

The enzyme concentration was 4 μmol dm^{-3} throughout the titration. Calculate the number of substrate binding sites on the enzyme and the dissociation constant, from these data.

2. Consider the binding of substrate (S) to enzyme (E). If an inhibitor, I, which can bind to the enzyme (but not to the ES complex) is now added to the system we have two simultaneous equilibria to be considered:

$$E + S \rightleftharpoons ES \qquad K_s = \frac{[E][S]}{[ES]},$$

---

† This must be done without perturbing the equilibrium (e.g. by using a dialysis membrane which is permeable to small molecules but not to macromolecules).

and

$$E + I \rightleftharpoons EI \qquad K_i = \frac{[E][I]}{[EI]}.$$

(a) Evaluate the concentrations of [E], [EI], and [ES] as fractions of $[E]_{total}$ for the case where

$$[E]_{total} = 1 \ \mu mol \ dm^{-3}, \qquad [S]_{total} = 150 \ \mu mol \ dm^{-3},$$
$$K_s = 50 \times 10^{-6} \ (mol \ dm^{-3}), \qquad K_i = 75 \times 10^{-6} \ (mol \ dm^{-3}).$$

(b) Discuss the situation (i.e. examine the algebra) when $[E]_{total}$ is comparable with $[S]_{total}$ and $[I]_{total}$ in this system.

3. If the binding of a ligand to a macromolecule is very tight, we can determine the number of ligand binding sites by a direct titration method. Such a case is provided by the binding of NADPH to isocitrate dehydrogenase from bovine heart mitochondria, which is monitored by the increase in fluorescence of NADPH when it is bound to the enzyme. The following data were obtained when NADPH was added to a $0·0135$ g dm^{-3} solution of the enzyme (a small correction has been made for the fluorescence of free NADPH). The molecular weight of the enzyme is 90 000.

NADPH added ($\mu mol \ dm^{-3}$)	0·07	0·14	0·21	0·28	0·35	0·50	0·75
Fluorescence increase		2·3	4·7	7·0	9·3	9·9	10	10

Determine the number of sites for NADPH on the enzyme.

4. The following data were obtained for the binding of Mn^{2+} to the enzyme phosphorylase a:

$[Mn^{2+}]_{total}(\mu mol \ dm^{-3})$	110	190	290	500	750
$[Mn^{2+}]_{bound}(\mu mol \ dm^{-3})$	75	125	175	250	300

The enzyme concentration is $100 \ \mu mol \ dm^{-3}$. By means of a suitable plot calculate n and K.

5. The binding of NAD$^+$ to yeast glyceraldehyde-3-phosphate dehydrogenase was studied by equilibrium dialysis with the following results:

$[NAD^+]_{total}(\mu mol \ dm^{-3})$	41	78	132	187	230	285	374	474
$[NAD^+]_{free} \ (\mu mol \ dm^{-3})$	13	21	30	39	48	68	125	211

The enzyme concentration is $71 \ \mu mol \ dm^{-3}$. What type of binding is being observed in this case?

6. The enzyme aspartate transcarbamoylase is inhibited by CTP. Measurements of the binding of CTP to the enzyme were made by equilibrium dialysis.

Experiment A

$[CTP]_{total} \ (\mu mol \ dm^{-3})$	0·78	1·29	2·35	4·63	11·65
$[CTP]_{free} \ (\mu mol \ dm^{-3})$	0·24	0·43	1·00	2·56	8·86

Experiment B

$[CTP]_{total} \ (\mu mol \ dm^{-3})$	144·9	172·0	203·7	269·5	416·0
$[CTP]_{free} \ (\mu mol \ dm^{-3})$	27·9	43·0	65·7	122·5	260·0

The concentrations of enzyme in experiments A and B were $0·27$ and 9 g dm^{-3} respectively. Aspartate transcarbamoylase has been shown to consist of six catalytic and six regulatory subunits with a total molecular weight of 300 000. What can you conclude from these data?

5. The thermodynamics of solutions

Introduction

IN Chapter 3 we noted that we could apply the equation $-\Delta G° = RT \ln K$ to reactions in solution, even though the equation was initially derived for reactions involving perfect gases. In this Chapter we shall show that there is a relationship between the properties of a solution and the vapour above the solution (Raoult's Law). This enables us to apply the equations derived for gaseous systems (in Chapter 3) to solutions. We shall also discuss some of the more important aspects of the thermodynamics of solutions. Initially we consider a one component system—a liquid in equilibrium with its vapour.

Equilibrium between phases

An obvious starting point to relate gases and liquids is to consider the vapour above a liquid in an isolated container (Fig. 5.1).

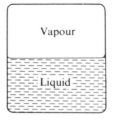

FIG. 5.1. Equilibrium between vapour and liquid in a closed container.

Under a given set of conditions (temperature, pressure) an equilibrium will be set up between the two phases. The condition for equilibrium between the phases is that

$$\Delta G = 0$$

i.e. $G_{\text{liquid}} = G_{\text{vapour}}.$

This means that the properties of the liquid (such as its free energy) can be related to those of its vapour, provided equilibrium conditions prevail.

Consider a simple example

$$\text{liquid} \rightleftharpoons \text{vapour}$$
$$(\text{pressure} = p_{\text{vap}}).$$

The equilibrium constant K is given by

$$K = \left(\frac{p_{vap}}{p_{vap}^\circ}\right),$$

where p_{vap}° refers to the standard state of the vapour, i.e. 1 atmosphere pressure.

The standard state of the liquid is simply the pure liquid, and we recall from Chapter 3 that components in their standard state (pure liquids, solids, gases at 1 atmosphere, or solutes at 1 mol dm^{-3} concentration) do not appear in the equilibrium constant expression.

The variation of K with temperature is given by the eqn (3.9)

$$\ln\left(\frac{K_2}{K_1}\right) = -\frac{\Delta H^\circ}{R}\left(\frac{1}{T_2} - \frac{1}{T_1}\right),$$

K can be replaced by p_{vap} at the appropriate temperature, $p_{vap}(T)$, so that

$$\ln\left(\frac{p_{vap}(T_2)}{p_{vap}(T_1)}\right) = -\frac{\Delta H^\circ}{R}\left(\frac{1}{T_2} - \frac{1}{T_1}\right),$$

where ΔH° is the difference in enthalpy between the liquid and vapour each in their standard state, i.e. $\Delta H^\circ =$ the latent heat of vaporization, L_{vap}.

$$\therefore\quad \boxed{\ln\left(\frac{p_{vap}(T_2)}{p_{vap}(T_1)}\right) = -\frac{L_{vap}}{R}\left(\frac{1}{T_2} - \frac{1}{T_1}\right).} \tag{5.1}$$

This is known as the *Clausius–Clapeyron* equation and shows how the vapour pressure of a liquid varies with temperature. A plot of ln (vapour pressure) against $1/T$ gives a straight line from which the latent heat of vaporization can be calculated. This equation provides a link between liquid and vapour phases since by making measurements of vapour pressure we can deduce the value of L_{vap}. (This reflects the forces between the molecules in the *liquid state*, since in the vapour there are no such intermolecular forces, if the vapour is assumed to be a perfect gas.)

Raoult's Law—ideal solutions

The most important equation linking vapour and liquid is known as *Raoult's Law*

$$\boxed{p = p^* X_1}, \tag{5.2}$$

where p is the vapour pressure of the solvent at temperature T, p^* is the vapour pressure of pure solvent at the same temperature and X_1 is the mole fraction of solvent

i.e. $X_1 = \dfrac{\text{number of moles of solvent}}{\text{number of moles of solvent} + \text{number of moles of solute}}.$

This relationship was initially based on experimental measurements of vapour pressure. An *ideal solution* can be defined as one which obeys Raoult's Law.† (In an analogous fashion, an *ideal gas* is defined as one which obeys the equation $PV = nRT$.) On a molecular level, an ideal solution implies that the intermolecular forces (which are responsible for the liquid state) are all uniform. Thus if the components are termed 1 and 2, the forces between 1 and 1, 2 and 2, and 1 and 2, are all identical. In an ideal gas, of course, there are no forces between the molecules.

The uniformity of the intermolecular forces in ideal solutions means that the enthalpy of mixing of the components is zero (i.e. $\Delta H = 0$) and suggests that such behaviour will be observed when the components are chemically very similar. A further implication is that there is no volume change on mixing the components.

Alternative definition of an ideal solution

Consider the vapour in equilibrium with a solvent.

Now for the vapour we have (Chapter 3) (eqn (3.1))

$$G_{\text{vap}} = G^\circ_{\text{vap}} + RT \ln p_{\text{vap}}, \quad \text{since } p^\circ_{\text{vap}} = 1 \text{ atm.} \tag{5.3}$$

To relate G°_{vap} to G°_{solvent} we have to consider the following cycle.‡

vapour at pressure p^*

(i) ↗   ↘ (ii)

(iii)
pure solvent → vapour at pressure 1 atm.

The ΔG for step $(i) = 0$ since pure solvent is in equilibrium with vapour at a pressure p^*.

For step (ii) (compression of vapour from pressure $= p^*$ to pressure $= 1$ atm.)

$$\Delta G = -RT \ln p^*/1 = -RT \ln p^*.$$

---

† If there is more than one volatile component in the solution, e.g. a solution of benzene in toluene, then Raoult's Law predicts the partial vapour pressure of each component. The total vapour pressure is then the sum of these partial vapour pressures. However we shall be dealing almost exclusively with solutions of non-volatile solutes where this complication does not arise.

‡ This type of approach can be used to solve problem 8, in Chapter 3.

For step (*iii*) ΔG is the sum of the ΔGs for steps (*i*) and (*ii*)

$$\text{i.e.}\quad \Delta G = -RT \ln p^*,$$

but since step (*iii*) is conversion of solvent in *its standard state* to vapour in *its standard state*, this is a $\Delta G°$

$$\text{i.e.}\quad \Delta G° = -RT \ln p^*,$$

$$\text{so that}\quad G°_{vap} - G°_{solvent} = -RT \ln p^*,$$

$$\therefore\quad \underline{G°_{vap} = G°_{solvent} - RT \ln p^*.} \tag{5.4}$$

Under any conditions where *equilibrium* between the solvent and its vapour exists, then

$$G_{solvent} = G_{vap},$$

substituting in eqn (5.3) this gives

$$G_{solvent} = G°_{vap} + RT \ln p_{vap}.$$

Using eqn (5.4)

$$G_{solvent} = G°_{solvent} - RT \ln p^* + RT \ln p_{vap}.$$

Now from *Raoult's Law*

$$p_{vap} = p^* X_1,$$

where X_1 is the mole fraction of solvent.

$$\therefore\quad G_{solvent} = G°_{solvent} - RT \ln p^* + RT \ln p^* + RT \ln X_1.$$

$$\therefore\quad \boxed{G_{solvent} = G°_{solvent} + RT \ln X_1}. \tag{5.5}$$

This fundamental equation can also be used to define an ideal solution,† i.e. the free energy of the solvent in an ideal solution is given by the expression above. We have shown how the equation follows from Raoult's Law.

We shall now use these complementary definitions of ideal solutions to discuss some of the properties of these solutions.

---

† Raoult's Law defines an ideal solution in terms of the properties of the *solvent*. There is another law (*Henry's Law*) dealing with the properties of the *solute* in an ideal solution. Henry's Law states the the mole fraction of the solute is *proportional* to the vapour pressure of the solute. Clearly this is of most use when the solute is volatile—e.g. oxygen gas dissolved in water. Since we are almost exclusively concerned with non-volatile solutes we shall not consider Henry's Law further.

Properties of ideal solutions

(1) *The lowering of the vapour pressure of the solvent*

Raoult's Law states

$$p = p^* X_1$$

$$\therefore \quad \frac{p^* - p}{p^*} = 1 - X_1$$

The left hand side of this equation is the relative lowering of the vapour pressure of the solvent caused by addition of solute.

The term on the right hand side is equal to the mole fraction of solute (X_2). X_2 is defined as

$$X_2 = \frac{\text{moles solute}}{\text{moles solute} + \text{moles solvent}}.$$

In *dilute solutions* the number of moles of solvent is very much greater than the number of moles of solute.

Thus

$$X_2 = \frac{\text{moles solute}}{\text{moles solvent}}.$$

If we have x g solute of molecular weight M in 1 kg of water (M.W. = 18)

$$X_2 = \frac{x/M}{1000/18} = \frac{x}{55 \cdot 5 \, M}$$

Now, from above

$$X_2 = \frac{p^* - p}{p^*}$$

$$\therefore \quad \boxed{\frac{p^* - p}{p^*} = \frac{x}{55 \cdot 5 \, M}}$$

So that if we measure the lowering of the vapour pressure of the solvent by addition of a known amount of solute, we can calculate the molecular weight of the solute.

(2) *Depression of the freezing point of the solvent*

It follows from the definition of an ideal solution (eqn 5.5),

$$G_{\text{solvent}} = G^\circ_{\text{solvent}} + RT \ln X_1$$

that the addition of solute has lowered the free energy of the solvent by an amount $RT \ln X_1$.† We shall now investigate the effect of addition of solute on the freezing point of the solvent.

---

† X_1, the mole fraction of solute is less than one, and so $\ln X_1$ is negative.

Consider the freezing of the solution

$$\text{solvent (liquid)} \rightleftharpoons \text{solvent (solid)}.$$

At the freezing point of the pure solvent (T_f), of course, $\Delta G = 0$ (this is an equilibrium situation). However on addition of solute, the free energy of the solvent (liquid) has been lowered by an amount $RT \ln X_1$. What we wish to find is the new temperature at which $\Delta G = 0$, when the solute is present. (This is the new freezing point of the solution.) Thus we need to use an equation describing the variation of ΔG with temperature. From eqn (3.6), we recall that this is

$$\frac{\mathrm{d}(\Delta G/T)}{\mathrm{d}T} = -\frac{\Delta H}{T^2}$$

Integrating between T_f, the freezing point of pure solvent (where $\Delta G = RT \ln X_1$) and T_f', the new freezing point (where $\Delta G = 0$), we have

$$\int_{\Delta G = RT \ln X_1}^{\Delta G = 0} \mathrm{d}\left(\frac{\Delta G}{T}\right) = -\int_{T_f}^{T_f'} \frac{\Delta H}{T^2}\, \mathrm{d}T$$

$$\therefore \quad -R \ln X_1 = +\Delta H\left(\frac{1}{T_f'} - \frac{1}{T_f}\right)$$

$$\therefore \quad -\ln X_1 = \frac{\Delta H}{R}\left(\frac{T_f - T_f'}{T_f' \times T_f}\right).$$

If $T_f \approx T_f'$ then we can set $T_f' \times T_f = (T_f)^2$ and X_1 can be replaced by $(1 - X_2)$ (where X_2 is the mole fraction of solute). We obtain

$$-\ln(1 - X_2) = \frac{\Delta H}{R} \cdot \frac{\Delta T_f}{T_f^2},$$

(where ΔT_f is the freezing point depression). If X_2 is small, $\ln(1 - X_2) \approx -X_2$

$$\therefore \quad X_2 = \frac{\Delta H}{R} \times \frac{\Delta T_f}{T_f^2}.$$

Now $\Delta H =$ the latent heat of fusion, L_f.

$$\therefore \quad \boxed{X_2 = \frac{L_f}{R} \times \frac{\Delta T_f}{T_f^2}}. \tag{5.6}$$

This is the equation linking the mole fraction of solute (X_2) with the freezing point depression (ΔT_f). Note that we have assumed that (a) the solution is ideal and (b) that X_2 is small, i.e. the solution is dilute.

By considering the effect of a solute on the free energy of the solvent in the equilibrium

$$\text{Solvent} \rightleftharpoons \text{vapour}$$

we can obtain an analogous expression for the boiling point elevation caused by a solute

$$X_2 = \frac{L_{\text{vap}}}{R} \times \frac{\Delta T_b}{T_b^2}$$

(5.7)

where L_{vap} is the latent heat of vaporization and T_b is the boiling point of pure solvent.

Now for the freezing point depression (eqn (5.6))

$$X_2 = \frac{L_f}{R} \times \frac{\Delta T_f}{T_f^2}$$

so that

$$\Delta T_f = \frac{RT_f^2}{L_f} \times X_2$$

substituting for X_2, as on page 49:

$$\Delta T_f = \frac{RT_f^2}{L_f} \times \frac{x}{55 \cdot 5 \, M}.$$

For water $L_f = 6 \cdot 02 \text{ kJ mol}^{-1}$, $T_f = 273 \cdot 15 \text{ K}$

$$\Delta T_f = 1 \cdot 86 \left(\frac{x}{M} \right)$$

so that if we had a 1 *molal* solution (1 mol solute per kg water) then

$$\Delta T_f = 1 \cdot 86 \text{ K}$$

This is called the *molal freezing point depression constant* for water. For other solvents its value is different, e.g. for benzene the constant is $5 \cdot 1$ K.

Worked example. A solution of 39 g glucose in 1 kg water has a freezing point $0 \cdot 4$ K below that of pure water. What is the molecular weight of glucose? The molal freezing point depression constant of water is $1 \cdot 86$ K.

Solution. Let the molecular weight of glucose $= M$. then the solution is $(39/M)$ molal.

$$\therefore \quad 0 \cdot 4 = 1 \cdot 86 \left(\frac{39}{M}\right)$$

$$\therefore \quad M = \frac{1 \cdot 86 \times 39}{0 \cdot 4}$$

$$= \underline{181}$$

i.e. the molecular weight of glucose is $\underline{181}$ (actual value $= 180$).

(3) *The osmotic pressure of a solution*

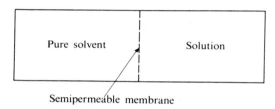

FIG. 5.2. Separation of solvent and solution by semipermeable membrane.

Consider a system in which a solution is separated from its pure solvent by a semipermeable membrane (i.e. the membrane is permeable to solvent, but not to solute) (Fig. 5.2).

Clearly the free energy of the solvent is different on the two sides of the membrane. On the solution side, the free energy of the solvent is given by eqn (5.5)

$$G_{\text{solvent}} = G^{\circ}_{\text{solvent}} + RT \ln X_1$$

whereas, on the solvent side, the free energy is $G^{\circ}_{\text{solvent}}$ (since we have pure solvent).

Since the free energies of the solvent on either side of the membrane are different, this is not an equilibrium situation, and there will be a tendency for solvent to pass into the solution.†

We can bring the system to equilibrium using various external factors. In the previous section we saw that changes in temperature could affect the value of G_{solvent}. Here we are concerned with the effects of pressure on the solution. The pressure required to bring the solution into equilibrium with

---

† As solvent passes into the solution, the mole fraction of solvent (X_1) in the solution tends towards 1 and hence G_{solvent} tends towards $G^{\circ}_{\text{solvent}}$.

the solvent is known as the *osmotic pressure* of the solution. This is illustrated in Fig. 5.3.

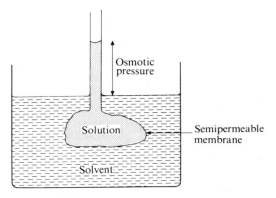

FIG. 5.3. The osmotic pressure of a solution separated from its solvent by a semipermeable membrane.

We shall derive the formula for osmotic pressure as follows:
  In Chapter 3 (p. 17) we saw that for any substance

$$dG = V\,dP - S\,dT$$

so that at constant temperature

$$dG = V\,dP.$$

Now the solvent on the solution side has a free energy $RT \ln X_1$ lower than that of pure solvent (this is because of dilution). The effect of applying the osmotic pressure must be to increase G_{solvent} in the solution (G_A) so as to counteract this dilution effect. At equilibrium (change in G from applying osmotic pressure) + (change in G from dilution) = 0.
  Let the osmotic pressure (in excess of 1 atm) = π atm, then

$$\int_0^{\pi} dG_A + RT \ln X_1 = 0$$

$$\int_0^{\pi} V_A\,dP = -RT \ln X_1.$$

V_A is the molar volume of solvent, and can be assumed to be independent of pressure (i.e. the solution is incompressible).

$$\therefore \quad V_A \pi = -RT \ln X_1.$$

As before X_1 is replaced by $(1 - X_2)$ and if X_2 is small (i.e. if the solution is dilute) we can write:

$$-RT \ln X_1 = -RT \ln(1 - X_2)$$

$$= RT \cdot X_2$$

$$\therefore \quad V_A \pi = RT \cdot X_2.$$

For *dilute aqueous* solutions, V_A is $0 \cdot 018 \, dm^3 \, mol^{-1}$ (since the specific gravity is close to unity) and the above equation becomes

$$\boxed{\pi = RT \cdot c} \qquad (5.8)$$

where c is the *molarity* of the solute (mol solute per dm^3 solution).

Worked example

What is the osmotic pressure of a solution containing 39 g glucose in $1 \, dm^3$ solution? ($T = 298 \, K$)

Solution

For dilute solutions

$$\pi = RTc.$$

If we measure π in atmospheres, we must use the appropriate value of R, i.e. $R = 0.082 \, dm^3 \, atm \, mol^{-1} \, K^{-1}$.

$$\therefore \quad \pi = (0 \cdot 082)(298)(c)$$

but $c = 39/180 \, mol \, dm^{-3}$

$$\therefore \quad \pi = (0 \cdot 082)(298)(39/180) \, atm$$

$$= 5 \cdot 3 \, atm.$$

The osmotic pressure is equivalent to a column of water 55 metres high!

This example shows that measurement of the osmotic pressure of a dilute solution is much easier than the measurement of the rather small freezing point depression ($0 \cdot 4 \, K$ in this case).

The four properties, vapour pressure lowering, freezing point depression, boiling point elevation, and osmotic pressure, are known as the *colligative properties* of solutions. They have been used for determination of the molecular weight of small molecules (solutes), but in dealing with solutions containing macromolecules, we shall see later that in general only osmotic pressure measurements are feasible.

Anomalous molecular weights

Sometimes the molecular weight appears to be 'anomalous'. This could arise from a number of factors.

(1) An electrolyte, e.g. NaCl, will be *dissociated* in solution. This increases the number of species present in the solution and the colligative properties will be greater than those predicted on the basis of the molar concentration of solute. The ratio (observed effect/predicted effect) is given the symbol i, and is known as the *van't Hoff factor* for the solution.

(2) In some cases the solute can *associate* in solution, e.g. benzoic acid can form dimers in aqueous solution.

$$2C_6H_5CO_2H \rightleftharpoons C_6H_5C \begin{smallmatrix} O\cdots H-O \\ \diagup \hspace{2em} \diagdown \\ \diagdown \hspace{2em} \diagup \\ O-H\cdots O \end{smallmatrix} C-C_6H_5$$

Mol. wt. 122   Mol. wt. 244

This reduces the number of species present in the solution, with a concomitant decrease in the colligative properties. The apparent molecular weight is between the monomer and dimer molecular weights depending on their relative numbers (concentrations).

(3) The solution may be non-ideal. This is likely to be particularly true when the solute and solvent are chemically dissimilar, such as in most of the solutions of interest to biochemists.

Non-ideal solutions—the concept of activity

From Raoult's Law we have been able to derive some simple and convenient equations to describe the properties of solutions. In practice the forces between molecules in a solution are not all uniform and there is a consequent departure from ideality. We would expect this departure to be small in dilute solutions, as indeed is the case. However, it remains convenient to keep the same form of the thermodynamic equations even for non-ideal solutions. To do this the term 'activity' is introduced as follows:

For ideal solutions Raoult's Law is

$$p = p^* X_1$$

while for non-ideal solutions this becomes

$$\boxed{p = p^* a_1}$$ (5.9)

where a_1 is the *activity* of the solvent. The thermodynamic equations are then exactly analogous if a_1 is substituted for X_1; a_1 is thus the mole fraction 'corrected' for the effects of non-ideality so that thermodynamic equations still hold.† The ratio of activity to mole fraction is termed the *activity coefficient* γ.

$$\gamma = \frac{a}{X} = \frac{\text{effective concentration}}{\text{true concentration}}.$$

(5.10)

When $\gamma = 1$, the solution is obviously ideal. Eqn. (5.10) enables us to calculate γ from a known value of X is a is measured.

Worked example

At 323 K, the vapour pressure of pure water is $0\cdot122$ atm. In a sucrose solution, where the mole fraction of water is $0\cdot96$, the vapour pressure is measured as $0\cdot1165$ atm (at 323 K). Calculate the activity coefficient of the water.

Solution

From Raoult's Law—for non-ideal solutions (eqn 5.9)

$$a_1 = \frac{p}{p^*} = \frac{0\cdot1165}{0\cdot122} = 0\cdot955$$

since

$$X_1 = 0\cdot96$$

we have

$$\gamma = \frac{0\cdot955}{0\cdot960} = \underline{0\cdot995}$$

i.e. in this dilute solution the solvent behaves almost ideally.

Electrolyte solutions

Clearly one case where the interactions between the species in solution are not going to be identical is that of solutions of electrolytes. By making the assumption that the departures from ideal behaviour in dilute solutions of electrolytes are a consequence of electrical interactions Debye and Hückel were able to derive the following expression for aqueous solutions:

---

† Thus eqn (3.1) $G = G° + nRT \ln (p/p^0)$ is replaced by $G = G^0 + RT \ln a$ and this equation is then assumed to apply to each component (gaseous, liquid, solute, etc.).

$$\boxed{\log \gamma_\pm = -0.5\, Z_+ Z_- \sqrt{I},} \;\dagger \qquad (5.11)$$

$\gamma_\pm$ is the mean activity coefficient of the ions. (The activity coefficient of the ions cannot be measured separately.) Z_+, Z_- are the charges carried by the positive and negative ions respectively. No account is taken of the sign of the charges. I is the *ionic strength*—defined as

$$\boxed{I = \tfrac{1}{2} \sum c_i Z_i^2.} \qquad (5.12)$$

Worked example

Calculate the mean activity coefficient in solutions of (a) 0.05 mol dm^{-3} MgCl$_2$ and (b) 0.1 mol dm^{-3} Na$_3$PO$_4$.

Solution

Both these salts are fully dissociated.

In (a) $Z_+ = 2$, $Z_- = 1$, $c_+ = 0.05$ mol dm^{-3} $c_- = (2 \times 0.05)$ mol dm^{-3} since there are two chloride ions per magnesium ion.

Thus

$$I = \tfrac{1}{2}(0.05 \times 2^2 + 2 \times 0.05 \times 1^2)$$
$$= 0.15.$$

From the Debye–Hückel equation

$$\log \gamma_\pm = -0.5 \times 2 \times 1 \times \sqrt{(0.15)}$$
$$\gamma_\pm = 0.41$$

In (b) $Z_+ = 1$, $Z_- = 3$, $c_+ = (3 \times 0.1)$ mol dm^{-3}, $c_- = 0.1$ mol dm^{-3}

Thus

$$I = \tfrac{1}{2}(0.3 \times 1^2 + 0.1 \times 3^2)$$
$$= 0.6.$$

Hence

$$\log \gamma_\pm = -0.5 \times 3 \times 1 \times \sqrt{(0.6)}$$
$$\gamma_\pm = 0.07.$$

---

† Eqn (5.11) is valid at 298 K. At other temperatures the constant (0·5) will be slightly different.

These mean ionic activity coefficients are related to the individual activity coefficients of the ions by the formula

$$(\gamma_\pm)^n = (\gamma_+)^{n^+}(\gamma_-)^{n^-}$$

where n is the *total* number of ions from one mole and $n+$ and $n-$ are the numbers of the individual ions. Thus for $MgCl_2$ this would become

$$(\gamma_\pm)^3 = (\gamma_{Mg^{2+}})^1 (\gamma_{Cl^-})^2$$

i.e.

$$\gamma_\pm = \{(\gamma_{Mg^{2+}})(\gamma_{Cl^-})^2\}^{\frac{1}{3}}$$

Similar expressions for the mean activities (i.e. $a_\pm$) and concentrations can also be written so that in the case of a *single salt*

$$\gamma_\pm = a_\pm / c_\pm$$

for the example of 0.05 mol dm^{-3} $MgCl_2$ $c_\pm$ is given by

$$c_\pm = \{(0.05)^1 (2 \times 0.05)^2\}^{\frac{1}{3}}$$

$$c_\pm = 0.079 \text{ mol dm}^{-3}$$

Thus

$$a_\pm = \gamma_\pm \cdot c_\pm$$

$$= 0.41 \times 0.079$$

$$a_\pm = 0.032 \text{ (mol dm}^{-3}).$$

This is the mean ionic activity of the solute that would be obtained from measurements of the solution properties.

Problem

What is the mean activity of 0.1 mol dm^{-3} Na_3PO_4?

Solution

$$a_\pm = 0.016 \text{ (mol dm}^{-3}).$$

The conclusion from the preceding discussions is that *activities* rather than *concentrations* should always be used in the expressions for the equilibrium constants. This is particularly true for solutions of electrolytes.† One example of this is afforded by the solubility of a sparingly soluble salt.

Sparingly soluble salts

Suppose a sparingly soluble salt such as $CaCO_3$ is added to water to give a saturated solution. For the equilibrium

$$CaCO_3 \text{ (solid)} \rightleftharpoons Ca^{2+}\text{(aqueous)} + CO_3^{2-} \text{ (aqueous)}$$

$$K_s = \frac{a_{Ca^{2+}} \times a_{CO_3^{2-}}}{a_{CaCO_3}}.$$

† In accurate work, extrapolations of data (e.g. equilibrium constants) are made to a solution of zero ionic strength. When $I = 0$, the solution behaves ideally, and thus 'true' thermodynamic constants can be obtained.

Since solid $CaCO_3$ is in its standard state, $a_{CaCO_3} = 1$.
   Thus

$$K_s = a_{Ca^{2+}} \times a_{CO_3^{2-}}.$$

K_s is known as the *solubility product* of the salt. Thus

$$K_s = (\gamma_+ c_+)(\gamma_- c_-)$$

$$= \gamma_\pm^2 c_+ \cdot c_-.$$

If $\gamma_\pm$ can be obtained from the Debye–Hückel law then K_s can be calculated
if c_+ and c_- are known.

Worked example

   AgCl has a solubility product of $1 \cdot 6 \times 10^{-10}$ (mol dm^{-3}) at 298 K. Calculate the concentration of Ag^+ ions in solution if $0 \cdot 01$ mol dm^{-3} NaCl is
present.

Solution

   The ionic strength of the solution is obviously entirely determined by the
NaCl which is present in large excess over the dissolved AgCl.
   Now

$$I = \tfrac{1}{2} \sum c_i Z_i^2$$

$$c_{Na^+} = c_{Cl^-} = 0 \cdot 01 \text{ mol dm}^{-3}$$

$$\therefore \quad I = \tfrac{1}{2}\{(0 \cdot 01)1^2 + (0 \cdot 01)1^2\}$$

$$= 0 \cdot 01.$$

From the Debye–Hückel expression

$$\log \gamma_\pm = -0 \cdot 5 \, Z_+ Z_- \sqrt{I}$$

$$\therefore \quad \gamma_\pm = 0 \cdot 891\dagger$$

Thus

$$K_s = a_{Ag^+} \cdot a_{Cl^-} = 1 \cdot 6 \times 10^{-10}$$

$$a_{Cl^-} = \gamma_{Cl^-} \cdot c_{Cl^-}$$

$$= 0 \cdot 891 \times 0 \cdot 01$$

$$= 8 \cdot 91 \times 10^{-3}$$

since [Cl$^-$] is essentially that of the NaCl.

$$\therefore \quad a_{Ag^+} = \frac{K_s}{a_{Cl^-}} = \frac{1 \cdot 6 \times 10^{-10}}{8 \cdot 91 \times 10^{-3}} = 1 \cdot 80 \times 10^{-8}$$

† Note that for this solution, where all the ions carry unit charge, $\gamma_\pm = \gamma_+ = \gamma_-$.

Hence

$$\gamma_{Ag^+} \cdot c_{Ag^+} = 1\cdot80 \times 10^{-8}$$

and

$$c_{Ag^+} = \frac{1\cdot80}{0\cdot891} \times 10^{-8}$$

$$= 2\cdot01 \times 10^{-8} \; (\text{mol dm}^{-3})$$

i.e. 2.01×10^{-8} mol dm^{-3} AgCl dissolves under these conditions.

In this example the effect of adding NaCl has been to depress the solubility of AgCl. This is because we have added a large excess of chloride ions. If we added a salt, e.g. NaNO$_3$ with no common ion, we would find that the solubility of AgCl would be increased. This arises because the high ionic strength would lower $\gamma_\pm$ and so $c_\pm$ would have to increase to compensate, so that K_s remains constant.

The concept of chemical potential

We have seen that reactions or systems at equilibrium are characterized by the condition that $\Delta G = 0$. If we consider a single component in two environments (as for example the solvent in an osmotic pressure experiment or the diffusible ions in the Donnan effect, p. 66) then the condition for equilibrium is that the free energy of this component is the same in the two environments. Clearly if they are not the same there will be a tendency to equalize them, e.g. by the 'flow' of solute from the high free energy environment to the low free energy environment (so that ΔG is negative). This can be likened to an electrical circuit where current flows from a region of *high* electrical potential to one of *low* electrical potential. The greater this difference in potential the greater this flow of current. In the same way, we can think of free energy as a *chemical potential*, μ. In fact the chemical potential of a component in a mixture is essentially equal to its molar Gibbs free energy. In describing solution properties many authors use 'chemical potential' rather than Gibbs free energy. We have not done so because the treatments are entirely equivalent and the introduction of chemical potential might lead to some confusion. However in the case of solutions of *ions*, it is more meaningful to use the term potential since we are dealing with charged species. We shall discuss the electrochemical potential of ions in Chapter 8.

PROBLEMS

1. On Pike's Peak (Colorado Springs) water boils at $364\cdot1$ K. Calculate the atmospheric pressure on Pike's Peak given that the latent heat of vapourization is $43\cdot6$ kJ mol^{-1} and that water boils at $373\cdot15$ K at 1 atm pressure.

2. At 295 K the vapour pressure of a solution of 10 g of a solute in $0 \cdot 1$ dm^3 water is $0 \cdot 0247$ atm. If the vapour pressure of pure water is $0 \cdot 025$ atm at the same temperature, calculate the molecular weight of the solute.
What difference would result if the solute is 50 per cent ionized according to the equation below?

$$AB \rightleftharpoons A^+ + B^-$$

3. At 310 K the vapour pressure of a 60 per cent solution (w/w) of glycerol in water† is $0 \cdot 0434$ atm. If the vapour pressure of water is $0 \cdot 0620$ atm at this temperature calculate the activity and activity coefficient of water in this solution (assume the vapour pressure of glycerol is negligible).

4. Calculate the freezing point depression if: (1) $0 \cdot 1$ g of ammonium sulphate (molecular weight 132) is dissolved in 10 g of distilled water; (2) $0 \cdot 1$ g of a protein of molecular weight 20 000 is dissolved in 10 g of distilled water; (3) a mixture of $0 \cdot 1$ g protein and $0 \cdot 1$ g ammonium sulphate is dissolved in 10 g of water. (The freezing point depression constant for a solution of 1 mol solute per 1 kg water is $1 \cdot 86$ K.) In the light of this result *comment* on the possibility of using this method to determine the molecular weight of a macromolecule.

5. Several insects are able to withstand prolonged exposure to low temperatures. In many cases the haemolymph (blood) of these insects contains large amounts of glycerol. What protection against freezing would be offered by the glycerol in the haemolymph of the parasitic wasp *Bracon cephi* which occurs at a concentration of approximately 30 per cent (w/w)? (The latent heat of fusion of water is $6 \cdot 01$ kJ mol^{-1}.)

6. An aqueous solution of glucose (molecular weight 180) has an osmotic pressure of $1 \cdot 55$ atm at 298 K. What is the freezing point depression of this solution? (Freezing point depression constant $= 1 \cdot 86$ K for 1 mol solute per kg water; $R = 0 \cdot 082$ dm^3 atm mol^{-1} K^{-1}.)

7. Calculate the ionic strength of
   (a) $0 \cdot 1$ mol dm^{-3} MgCl$_2$.
   (b) $0 \cdot 05$ mol dm^{-3} Na$_2$HPO$_4$.
   (c) A 1 mmol dm^{-3} protein solution (as its sodium salt) at a pH at which the protein carries a charge of -10 units.
   (d) A $0 \cdot 1$ mol dm^{-3} solution of acetic acid ($K_a = 1 \cdot 75 \times 10^{-5}$ (mol dm^{-3})): see Chapter 7).

8. State the Debye–Hückel Law for a dilute aqueous solution of a strong electrolyte at 298 K.
   Calculate the activities of sodium and chloride ions in an aqueous solution of 1 mmol dm^{-3} sodium chloride, and also in an aqueous solution containing 1 mmol dm^{-3} sodium chloride and 3 mmol dm^{-3} potassium bicarbonate.

9. The total concentration of ferric ion in the plasma is estimated to be 50 μmol dm^{-3}. Calculate the pH at which 99 per cent of the Fe^{3+} would be precipitated, given that the solubility product for Fe(OH)$_3$ is 10^{-36} at 310 K. Assume that the ionic product, K_w, for water is 10^{-14} (K_w is the product of [H$^+$] and [OH$^-$], both expressed as mol dm^{-3}).

---

† i.e. the mixture contains 60 per cent (by weight) glycerol and 40 per cent water.

Comment on the state of Fe^{3+} in the plasma where the $[H^+]$ is about 10^{-7} mol dm^{-3}.

10. The mean standard enthalpy and entropy changes in the range 275 K to 298 K for the dissolution of $Ca(OH)_2$ in water to give an ideal solution of its constituents are $-16 \cdot 3$ kJ mol^{-1} and -153 J K^{-1} mol^{-1} respectively. Calculate the ideal solubility of $Ca(OH)_2$ at 298 K and 275 K (atomic weights Ca = 40, H = 1, O = 16).

11. The solubility product of AgCl is $1 \cdot 6 \times 10^{-10}$ (mol dm^{-3}) at 298 K. Calculate the concentration of Ag^+ ions in solution, if $0 \cdot 01$ mol dm^{-3} $NaNO_3$ is present. Compare your answer with that of the worked example on page 59 and comment on the difference.

6. Solutions of macromolecules

Introduction

IN Chapter 5 we discussed the thermodynamics of low molecular weight solutes. Clearly a large number of biologically interesting molecules are of high molecular weight and these are termed macromolecules, e.g. proteins, nucleic acids, polysaccharides, etc. The discussion of the thermodynamic properties of these molecules introduces several new features. One of these new features is the idea that the solution can contain species of different molecular weights and so average molecular weights can be involved. (We shall discuss this topic later.) Secondly, solutions of macromolecules would not be expected to be ideal because of the obvious chemical differences between solvent and solute. It is extremely difficult to calculate the activity coefficients in solutions of macromolecules, since there is no simple theory available (as there is, for instance, in the case of strong electrolytes). However, as before, in extremely dilute solutions ideal behaviour will be approached. Thus in order to apply the thermodynamic equations to solutions of macromolecules, it is necessary to extrapolate the results of any measurements to infinite dilution.

We perform the extrapolations using empirical equations based on the thermodynamic ones already described for ideal solutions. This is illustrated by considering the osmotic pressure of solutions of macromolecules.

For dilute ideal solutions we derived the relationship

$$\pi = RTc$$

where c is the molar concentration of solute. If the concentration of solute is x g dm^{-3} and the molecular weight is M, then

$$c = x/M$$

hence

$$\frac{\pi}{xRT} = \frac{1}{M}.$$

Now for solutions of macromolecules the equation above is inadequate and is replaced by a more complex one

$$\frac{\pi}{RTx} = \frac{1}{M} + \frac{B(x)}{M} + \frac{C(x^2)}{M^2} + \cdots$$

where B, C, etc. are constants known as *virial coefficients*.[†] Since M and M^2

[†] We should note the analogy with the equation often used to describe a non-ideal gas i.e. $P/RT = 1/V + BP + CP^2 + \dots.$

are also constants we can write:

$$\frac{\pi}{RTx} = \frac{1}{M} + B'(x) + C'(x)^2 + \cdots$$

where B', C' are new constants.

If x/M is small (i.e. the solution is dilute) then $(x/M)^2$ and higher terms are very small compared with x/M, and so we obtain

$$\boxed{\frac{\pi}{RTx} = \frac{1}{M} + B'(x)} . \qquad (6.1)$$

Since the first term on the right hand side of this equation is a constant, a plot of π/RTx against x is a straight line. The intercept on the y-axis can be used to calculate the molecular weight of the solute:

$$\left(\frac{\pi}{RTx}\right)_{x=0} = \frac{1}{M}$$

$$\therefore \quad M = \frac{1}{(\text{Intercept on } y\text{-axis})}.$$

Worked example

The following osmotic pressure data were obtained for solutions of bovine serum albumen at 298 K.

Protein concentration (g dm^{-3})	18	30	50	56
Osmotic pressure (atm)	0·0074	0·0134	0·0253	0·0292

Calculate the molecular weight of bovine serum albumen.

Solution

We evaluate π/xRT for each value of x. Note that π is in atmospheres and $R = 0.082$ dm^3 atm mol^{-1} K^{-1}.

x (g dm^{-3})	18	30	50	56
π/xRT ($\times 10^5$)	1·68	1·83	2·07	2·13

From the plot of π/xRT against x (Fig. 6.1), the intercept on the y-axis is 1.47×10^{-5}. Hence the molecular weight of bovine serum albumen is 68 000.

We should note that a plot of π against x is not linear (Fig. 6.2), thus showing the necessity to use the expanded equation for osmotic pressure

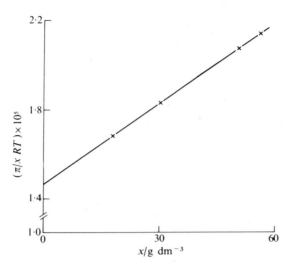

FIG. 6.1. Plot of osmotic pressure data from 'Worked example' according to eqn (6.1).

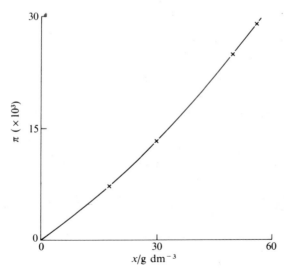

FIG. 6.2. Plot of osmotic pressure against concentration. (Data from the 'Worked example'.)

and performing the extrapolation to infinite dilution (Fig. 6.1).† For the case of an ideal solution there would be no change in the value of the term π/xRT as x varies.

The determination of the molecular weight of a macromolecule from the other colligative properties of the solution (i.e. vapour pressure lowering and freezing point depression) is not feasible because the effects observed are so small. This point is illustrated in Problem 4 of Chapter 5.

Significant exceptions to this statement have been found in the cases of the sera of fish living in cold waters (e.g. Arctic or Antarctic regions). These sera contain (in addition to small solutes) special types of proteins, known as 'anti-freeze proteins', which can exert a much larger effect on the freezing point of the sera than would be expected on the basis of their molecular weights. For instance, the average temperature in McMurdo Sound in Antarctica is 1·85 K below the freezing point of pure water and sera from fish living in temperate zones would freeze at this temperature, since they do not contain 'anti-freeze' proteins. However, in two species of Antarctic fish the serum freezing point was 2 K below that of pure water, enabling these fish to survive in these conditions. Investigations of the structures of these anti-freeze proteins have shown that they are based on relatively simple repeating units of small numbers of amino acids (often containing large quantities of attached carbohydrate). The mechanism by which this type of structure exerts its effect is under current investigation.

Apart from the non-ideality of solutions of macromolecules, we often have to consider the effects associated with macromolecules which carry a net charge. A very important example of this is the *Donnan* effect.

The Donnan effect

Consider two solutions separated by a membrane permeable to small ions but not to charged macromolecules (e.g. a dialysis membrane is not permeable to proteins). If the charged protein is present on only one side of the membrane, there will be an imbalance of ions across the membrane at equilibrium. This is known as the *Donnan effect*. (Sometimes it is also called the *Gibbs–Donnan effect*.)

For simplicity consider two compartments of equal volume, separated by the membrane (Fig. 6.3). Suppose we add an x molar solution of protein (with a charge of Z) and its associated ion (e.g. Na^+) to one side of the membrane, and that we add x_2 molar NaCl to the other side. Movement of ions will occur across the membrane but since electrical neutrality must be preserved at all times on both sides of the membrane, it follows that if y g ions of Na^+ move from side 2 to side 1, y g ions of Cl^- must also migrate in this direction. Equilibrium is achieved when the free energy of all species, to which the membrane is permeable is the same on either side of the

---

† Similar extrapolation procedures are used in the analysis of sedimentation and viscosity data of solutions of macromolecules.

Initial state   Final state

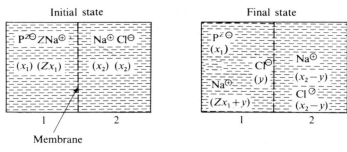

FIG. 6.3. Movement of ions across a membrane on one side of which charged macromolecules are present.

membrane. In this case we consider NaCl and so, at *equilibrium*

$$(G_{NaCl})_1 = (G_{NaCl})_2$$

(the subscripts refer to the appropriate side of the membrane). Using the definition (see footnote, p. 56)

$$G = G° + RT \ln a \qquad (a = \text{activity})$$

we obtain

$$(a_{NaCl})_1 = (a_{NaCl})_2$$

or†

$$(a_{Na^+} \cdot a_{Cl^-})_1 = (a_{Na^+} \cdot a_{Cl^-})_2$$

at equilibrium. If y moles of NaCl have been transferred when equilibrium is reached, then this condition becomes

$$(Zx_1 + y)(y) = (x_2 - y)(x_2 - y)$$

provided the activity can be replaced by concentration. From this equation

$$y = \frac{x_2^2}{Zx_1 + 2x_2} \tag{6.2}$$

and we can calculate the final composition of the solution on either side, if the values of x_1, x_2, and Z are known.

The ratio of the Cl⁻ ion concentration on side 2 to that on side 1 at equilibrium is $(x_2 - y)/y$, and using the above value for y, this ratio becomes $(Z_1 x_1 + x_2)/x_2$ or more explicitly:

---

† Note that $a_{NaCl} = a_{Na^+} \cdot a_{Cl^-}$. This follows since $G = G° + RT \ln a$ for each ion.

$$\left[\frac{(Cl^-)_2}{(Cl^-)_1}\right]_{(final)} = 1 + \left[\frac{(Na^+)_1}{(Na^+)_2}\right]_{(initial)}. \tag{6.3}$$

Worked example

At a certain pH ribonuclease carries a net negative charge of 3. If $0 \cdot 1$ dm^3 of a solution of 3 mmol dm^{-3} ribonuclease (as its sodium salt) is added to one side of a membrane, and $0 \cdot 1$ dm^3 of 50 mmol dm^{-3} NaCl is added to the other side, what are the final concentrations of ions on the two sides of the membrane?

Solution

The concentration of Na$^+$ ions initially present on the protein side is 9 mmol dm^{-3}.

Let y mmol dm^{-3} NaCl be transferred to the protein side at equilibrium. Then using the formula (eqn 6.2)

$$y = \frac{x_2^2}{Zx_1 + 2x_2}$$

$$= \frac{50 \times 50}{9 + 100} \text{ mmol dm}^{-3}$$

(since $\quad x_2 = 50$ mmol dm^{-3}, $Z = 3$, $x_1 = 3$ mmol dm^{-3})

hence $\quad y = 22 \cdot 94$ mmol dm^{-3}

so that on the protein side we have (at equilibrium)

$$[\text{Protein}^{3-}] = \underline{3 \text{ mmol dm}^{-3}}, \quad [\text{Na}^+] = \underline{31 \cdot 94 \text{ mmol dm}^{-3}},$$

$$[\text{Cl}^-] = \underline{22 \cdot 94 \text{ mmol dm}^{-3}}$$

and on the non-protein side,

$$[\text{Na}^+] = [\text{Cl}^-] = (50 - 22 \cdot 94) \text{ mmol dm}^{-3}$$

$$= \underline{27 \cdot 06 \text{ mmol dm}^{-3}}.$$

Problem

How would the result be different if we initially had $0 \cdot 2$ mol dm^{-3} NaCl on the non protein side?

Solution

protein side: $[\text{Protein}^{3-}] = \underline{3 \text{ mmol dm}^{-3}}$, $[\text{Na}^+] = \underline{106 \cdot 8 \text{ mmol dm}^{-3}}$, $[\text{Cl}^-] = \underline{97 \cdot 8 \text{ mmol dm}^{-3}}$; non-protein side: $[\text{Na}^+] = [\text{Cl}^-] = \underline{102 \cdot 2}$ $\underline{\text{mmol dm}^{-3}}$.

Studies of the binding of small charged ligands to macromolecules are often performed by a technique known as *equilibrium dialysis*. A solution containing the macromolecule is placed on one side of a membrane, and a solution of the ligand is added to the other side. When equilibrium has been attained,† the concentrations of ligand on the two sides of the membrane are measured. On the side containing the macromolecule the ligand concentration is the sum of free ligand and bound ligand. These data are analysed by the methods discussed in Chapter 4.

In any such study we should consider the Donnan effect, since this can also lead to differences in the concentrations of charged ligands between the two sides of the membrane. It is thus important, in binding studies to minimize the Donnan effect. For instance, in our example, we see from eqn (6.3) that at equilibrium the ratio $(Cl^-)_2/(Cl^-)_1$ approaches unity when the term $[(Na^+)_1/(Na^+)_2]_{initial}$ is small. Thus Donnan effects can be minimized by: (1) lowering the macromolecular concentration (so that $(Na^+)_1$ initial is small), (2) working at high ionic strengths (so that $(Na^+)_2$ initial is large), or (3) adjusting the pH so that the protein carries no net charge (i.e. it is at its *isoelectric* point—see Chapter 7). We should also take these precautions when doing osmotic pressure measurements on solutions of macromolecules, since the differences in ionic concentrations themselves give rise to an osmotic pressure.

The molecular weights of macromolecules

So far, we have made the assumption that there is a unique molecular weight for a macromolecule. This assumption is justified if there is only one type of species present as in the case of most purified proteins. Such solutions are termed *monodisperse* (see Fig. 6.4(*a*)) and the definition of molecular weight is self explanatory. However, in many instances solutions of biological macromolecules contain species with various molecular weights. If the molecular weight distribution is continuous (e.g. Fig. 6.4(*b*)),

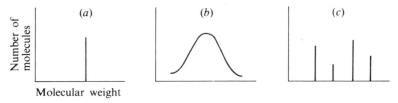

FIG. 6.4. Graphs depicting the molecular weight distributions for (a) monodisperse, (b) polydisperse, and (c) paucidisperse systems.

† That is, when the concentration of ligand on either side shows no further tendency to change.

the system is termed *polydisperse*. (This might be the case, for example, for nucleic acid fragments.) If the distribution of molecular weight is discrete e.g. Fig. 6.4(*c*)), the system is termed *paucidisperse*. An association–dissociation equilibrium would be an example of a paucidisperse system.

It is useful to define certain average molecular weights to describe *polydisperse* and *paucidisperse* systems.

Number-average molecular weight ($\bar{M}_n$)

In this average, the species present in solution contribute according to their relative populations so that the average is defined by

$$\bar{M}_n = \frac{\sum n_i m_i}{\sum n_i} \qquad (6.4)$$

where n_i is the relative number of the ith species, and m_i is its molecular weight.

Consider the example of enolase. At a certain concentration suppose that 60 per cent of the species are monomers of molecular weight 41 000 and 40 per cent are dimers of molecular weight 82 000.

The number-average molecular weight is then given by:

$$\bar{M}_n = \frac{\dfrac{60}{100}(41\ 000) + \dfrac{40}{100}(82\ 000)}{\dfrac{60}{100} + \dfrac{40}{100}} = \underline{57\ 400}.$$

This molecular weight would be obtained from measurements of colligative properties (e.g. osmotic pressure) since these properties depend only on the concentration of the species present, irrespective of size.[†] We shall see later that it can also be derived from sedimentation equilibrium.

Weight-average molecular weight ($\bar{M}_w$)

In this average the species present in solution contribute according to their relative weights, so that this average is defined by:

$$\bar{M}_w = \frac{\sum n_i m_i^2}{\sum n_i m_i} \quad .[‡] \qquad (6.5)$$

[†] The number-average molecular weight is also obtained by those techniques which 'count' the numbers of molecules directly (e.g. analysis of end groups, or active sites).

[‡] If we note that $x_i \propto n_i m_i$, where x_i is the *concentration* of the ith species (in g dm^{-3}) then the formula becomes:
$$\bar{M}_w = \frac{\sum x_i M_i}{\sum x_i}.$$

Thus in the above case of enolase, the weight average molecular weight is given by:

$$\bar{M}_w = \frac{\dfrac{60}{100}(41\,000)^2 + \dfrac{40}{100}(82\,000)^2}{\dfrac{60}{100}(41\,000) + \dfrac{40}{100}(82\,000)} = \underline{64\,430}.$$

This average molecular weight is larger than the number average because it gives increased importance to the heavier species. We could obtain $\bar{M}_w$ from studies of light scattering, sedimentation equilibrium, etc.

Z-Average molecular weight

In this average the species present in solution contribute according to the square of their relative weights, so that this average is defined by:

$$\boxed{\bar{M}_z = \frac{\sum n_i m_i^3}{\sum n_i m_i^2}}. \tag{6.6}$$

Thus in the above example of enolase, the z-average molecular weight can be shown to be 70 820.

The z-average molecular weight is easily obtained from sedimentation equilibrium measurements. Although the z-average molecular weight does not have a simple physical meaning, it can be used in conjunction with the other molecular weight averages to provide information on the composition of a *poly-* or *paucidisperse* system. (Clearly for a *monodisperse* system all the molecular weight averages will, of course, be the same.)

Determination of molecular weights

Of the many methods that can be used to determine molecular weights, we shall concentrate on the three that are probably the most widely used. Two of these (electrophoresis and gel filtration) are *semi-empirical* in that they are based on calibrations with standards of known molecular weight. The third method (analytical ultracentrifugation) is an *absolute* method, i.e. does not depend on comparisons with molecular weight standards. Because the analytical ultracentrifuge provides the only direct measure of molecular weights, we shall discuss its use in some detail. The section dealing with this is more advanced than much of this chapter and may be omitted at a first reading. Nevertheless the importance of this technique justifies its inclusion here.

Semi-empirical methods for the estimation of molecular weights

Two of the most popular techniques currently in use for molecular weight determination are electrophoresis and gel filtration. (Their popularity may be due in large measure to the fact that no expensive apparatus is required and that no complex mathematical equations are involved!)

Electrophoresis

Electrophoresis is the name given to the movement of a charged molecule under the influence of an applied electric field. In order to minimize the effects of convection, electrophoresis is normally conducted on a support material such as paper or more commonly a cross-linked gel composed of polyacrylamide. The rate of movement, or *mobility*, of a molecule in an electric field of given strength will depend on a number of factors such as the charge, size, and shape of the molecule and the nature of the support medium. Clearly, the method cannot be used to estimate the molecular weight of a molecule unless comparison is being made with some other species which is known to be similar. However, this limitation can be neatly overcome in the case of proteins by performing the gel electrophoresis in the presence of a detergent, sodium dodecyl sulphate (SDS—$CH_3(CH_2)_{11}OSO_3^-Na^+$) which breaks the non-covalent bonds maintaining the shape of the protein. It has been found that (i) nearly all proteins bind a constant amount of SDS per g protein (the negative charge of SDS overwhelms the intrinsic charge of the protein), and (ii) protein–SDS complexes are rod-shaped: the length of the rod depending on the molecular weight of the protein. Since the charge and hydrodynamic properties are thus merely functions of the molecular weight of the protein, it is not difficult to see why a relationship between mobility and molecular weight should exist. It is found in practice that a graph of log (molecular weight) vs. mobility is linear over a wide range of molecular weights of standard proteins. The molecular weight of the protein of interest can then be read off the standard line. Using gels of different acrylamide concentrations the range of molecular weights can be varied: low per cent acrylamide gels are used in the higher molecular weight ranges.

An analogous approach can be used for determination of the molecular weight of a sample of RNA. In this case formamide is added to destroy the secondary structure of the molecule (base pairing and stacking) and a graph of log (molecular weight) against mobility is linear over a wide range of molecular weights. Using this approach, RNA samples of known molecular weight were employed to construct a standard curve and then it was shown that haemoglobin messenger RNA had a molecular weight of about 200 000.

Gel filtration

In this method a cross-linked polymer in bead form is used to separate molecules according to size. If a solution is passed through a column (or over a plate) containing the polymer small molecules enter the pores of the bead and are retarded relative to larger molecules which cannot enter the pores and thus pass straight through the column. Molecules of intermediate size will be retarded to an intermediate extent. The usual procedure is to determine the elution volumes for a number of proteins of known molecular weight and construct a standard curve (elution volume against log (molecular weight)) from which the molecular weight of the molecule of interest can be determined. By varying the degree of cross-linking of the polymer, the range of molecular weights in the 'fractionation range' can be altered, e.g. Sephadex G-200 (a cross-linked polysaccharide) is useful in the range of molecular weights from 20 000 to 300 000.

The success of the method depends on the various molecules having roughly similar shapes. Globular proteins are very suitable in this connection, but proteins of a different shape (e.g. myosin which is rod-shaped) would not be expected to fall on the standard curve. Recent work, for instance, has shown that glycoproteins give anomalously high molecular weights in this method. The gel filtration technique is suitable for nucleic acid work only with the smaller nucleic acids (e.g. 4S and 5S RNA).

The analytical ultracentrifuge

In a closed, isolated system the temperature, pressure, and the concentration of all components are the same in all parts of the system. For a two component system we may designate one component as solvent and the concentration, c, of the solute will be the only composition variable.

If we introduce an external force (gravity, centrifugal) which acts only on the mass of any part of the system and is directly proportional to the mass, the uniform concentration distribution will be disturbed. Those macromolecules which are more dense than the solution will sediment and those which are less dense will move upwards until a new equilibrium is reached.

In the analytical ultracentrifuge the sample is enclosed in a transparent cell and can be considered as a closed, isolated system in a strong and well defined external, centrifugal field. The temperature is kept at a fixed, constant value and the concentration distribution can be monitored by optical methods.

The optical methods are normally based on the measurement of the refractive index as a function of position within the cell (schlieren and interferometer optics) or on the absorption properties of the solute (photoelectric scanner).

Sedimentation velocity

If the ultracentrifuge is operated at relatively high speeds (up to 65 000 r.p.m.†), the uniformly distributed macromolecules will migrate

---

† In a typical ultracentrifuge this would generate a centrifugal field of the order of 300 000 times gravity!

toward the bottom of the cell leaving a region which contains only solvent molecules, and creating a boundary between the solvent and the solution. *The sedimentation velocity method is essentially based on the measurement of the velocity of movement v of this boundary*, and the sedimentation coefficient of the macromolecule can be determined. The sedimentation coefficient, s, is defined by:

$$s = \frac{\mathrm{d}r/\mathrm{d}t}{\omega^2 r} = \frac{v}{\omega^2 r}$$

where ω is the angular velocity (radians s^{-1}) of the rotor and r is the distance of the boundary, measured from the axis of rotation.

Numerically, the *sedimentation coefficient* is the distance in cm, covered by a macromolecule during 1 s under the effect of a 'force' of 10^{-2} N kg^{-1}, in H_2O solvent at 293 K. Sedimentation coefficients are usually given in Svedberg, S, units; $1/S = 10^{-13}$ s.

The sedimenting macromolecule in the cell of an ultracentrifuge moves under the effect of three forces.

(i) The centrifugal force, F_c

$$F_c = \omega^2 r M$$

where M is the molecular mass.

(ii) The buoyant force, F_b

$$F_b = -\omega^2 r M \bar{v} \rho$$

where $\bar{v}$ is the partial specific volume of the molecule † and ρ is the density of the solvent. This force (which is analogous to the 'upthrust' of the Archimedes principle) opposes the centrifugal force because of displacement of solvent by the particle.

(iii) the frictional force, F

$$F = -fv$$

where f is the frictional coefficient (which depends on the shape of the particle and on the viscosity of the solvent) and v is the velocity of the particle relative to the solvent. This force opposes the motion.

Initially the particle will accelerate and reach a certain velocity at which the total force is zero. The particle will then sediment with this velocity v,

---

† In the case of a solution the total volume is not the sum of the volumes of the solute and solvent. The volume change on addition of solute depends on the nature of both the solution and the solvent. This leads to the definition of partial specific volume ($\bar{v}$) of a solute, in the case of a particular solvent, as *the volume change upon addition of 1 g of solute to a large volume of solution keeping all other parameters (temperature, pressure, etc.) constant.*

given by

$$\omega^2 rM - \omega^2 rM\bar{v}\rho - fv = 0,$$

rearranging and substituting

$$\boxed{\frac{M(1 - \bar{v}\rho)}{f}} = \boxed{\frac{v}{\omega^2 r}} = s. \qquad (6.7)$$

Molecular   Measurable
parameters   parameters

Eqn (6.7) expresses the sedimentation coefficient in terms of molecular parameters on the one hand and in terms of measurable quantities on the other. It is apparent from eqn (6.7) that the sedimentation coefficient depends on

(a) molecular properties of the sedimenting particle, such as: molecular weight, M, partial specific volume $\bar{v}$ and shape, f.

(b) properties of the solution: density, ρ and viscosity η (since f depends on this).

Because of the dependence of the sedimentation coefficient on the properties of the solution, the measured s values should be corrected to standard conditions to obtain comparable values for different molecules. The standard solvent is H_2O at 293 K and the s value under these conditions is denoted by $s_{20,w}^{\circ}$.†

For macromolecular solutions which are generally non-ideal, s should be measured at four or five different concentrations, and the corrected $s_{20,w}$ values should be extrapolated to zero concentration, to obtain the standard $s_{20,w}^{\circ}$ value, a molecular parameter, which reflects the molecular weight, shape, and density of a macromolecule. An experimentally more convenient form of eqn (6.7) can be obtained by integration‡ (noting that $v = dr/dt$)

$$\boxed{s = \frac{1}{\omega^2} \frac{d \ln r}{dt}}. \qquad (6.8)$$

The radial distance of the boundary is measured at various values of elapsed time during the run and a plot of $\ln r$ vs. t gives a straight line, the

---

† Strictly speaking this value should be written $s_{293,w}^{\circ}$ in SI units. However, the use of $s_{20,w}^{\circ}$ is so widespread that we retain it here.

‡ Since the velocity of the boundary is $v = dr/dt$, then $v = dr/dt = r\omega^2 s$. Rearranging $dr/r = \omega^2 s\, dt$ which gives on integration $\ln r(t)/r(t_0) = \omega^2 s(t - t_0)$ where $r(t)$ is the position of the boundary at time t.

slope of which can be used to compute the sedimentation coefficient. ω can be calculated from the rotor speed

$$\omega = \frac{2\pi \ \text{r.p.m.}}{60} = 0.1047 \ \text{r.p.m.}$$

If r is measured in cm and t in seconds s will be obtained in seconds(s). The value of s from eqn (6.8) is divided by 10^{-13} to obtain sedimentation coefficients in Svedberg units, S.

It should be stressed that the value of the sedimentation coefficient of a molecule is not sufficient to estimate its molecular weight. It can be used to identify macromolecular components of known molecular weight and conformation or to check the homogeneity of the sample. If more than one macromolecular component is present, this will be reflected in the sedimentation velocity schlieren patterns either as separate peaks or as an asymmetric single peak.

Table 6.1 presents sedimentation coefficient and molecular weight data for some macromolecules. Note the differences in molecular weight of molecules with similar sedimentation coefficients.

TABLE 6.1

Sedimentation coefficient and molecular weight data for selected macromolecules

	$s_{20,w}^{\circ}(S)$	Molecular weight
Myosin	6.4	540 000
α-Amylase	4.5	52 000
Fibrinogen	7.9	330 000
Glyceraldehyde-3-phosphate dehydrogenase	7.9	146 000
Ribonuclease	1.64	12 400
Bushy stunt virus	132	10 700 000

In sedimentation velocity experiments a boundary is created between the solution and the solvent. The concentration gradient is the driving force for diffusion which tends to spread the boundary. The flow of matter through a unit surface, driven by a concentration gradient (given by Fick's first law), is equal to the concentration gradient multiplied by the diffusion coefficient, D. D is in units of $\text{cm}^2 \ \text{s}^{-1}$ (or $\text{m}^2 \ \text{s}^{-1}$).

For ideal solutions $D = RT/f$, where f is the frictional coefficient. Substituting this into eqn (6.7) the molecular weight can be expressed in terms of the sedimentation (s) and diffusion (D) coefficients:

$$M = \frac{RTs}{D(1 - \bar{v}\rho)}.$$ (6.9)

This is the *Svedberg equation*, which is commonly used to calculate the molecular weight from the experimentally determined parameters s and D. it should be noted that in a *poly-* or *paucidisperse* system the molecular weight obtained from eqn (6.9) is not any of the averages discussed earlier (M_n, M_w, M_z). It is, in fact, a *mixed average* (denoted by M_{sD}).

The validity of the Svedberg equation is restricted to two component systems at *infinite dilutions*. For many globular molecules $s_{20,w}^{\circ}$ can be directly measured in very dilute solutions ($c < 0.1$ per cent) at 293 K in dilute aqueous buffers. If the values of both the diffusion coefficients and partial specific volume at 293 K are also known, the molecular weight can be calculated.

Worked examples

(1) Calculate the molecular weight of D-glyceraldehyde-3-phosphate dehydrogenase from the following data:

$$s_{20,w}^{\circ} = 7.865 \text{ S}; \quad D_{293,w}^{\circ} = 4.75 \times 10^{-11} \text{ m}^2 \text{ s}^{-1} \quad \bar{v}_{293} = 0.729 \text{ cm}^3 \text{ g}^{-1}.$$

Solution

Use the Svedberg equation

$$M_{sD} = \frac{RTs}{D(1 - \bar{v}\rho)}$$

$R = 8.31 \text{ J K}^{-1} \text{ mol}^{-1}; \quad \rho = 0.9982 \text{ kg dm}^{-3} \text{ (water at 293 K) } T = 293 \text{ K}$

$$s = 7.86 \times 10^{-13} \text{s}; \quad D = 4.75 \times 10^{-11} \text{ m}^2\text{s}^{-1}$$

$$M_{sD} = \frac{8.31 \times 293 \times 7.86 \times 10^{-13}}{4.75 \times 10^{-11}(1 - 0.729 \times 0.9982)} \text{ kg mol}^{-1}$$

$$= 148.0 \text{ kg mol}^{-1},$$

i.e.

$$M_{sD} = 148\ 000 \text{ g mol}^{-1}$$

The molecular weight is thus <u>148 000.</u>

In certain cases the experiments to determine the sedimentation coefficient cannot be conducted at 293 K or near infinite dilution. In such cases the obtained *apparent sedimentation coefficient values have to be corrected to standard conditions* (293 K in water) and extrapolated to zero concentration before using them either for comparison or to calculate the molecular weight.

(2) In order to preserve the enzyme an ultracentrifuge run was carried out on D-glyceraldehyde-3-phosphate dehydrogenase at 278 K. Only schlieren optics were available on the ultracentrifuge (concentration range 0·1–1·0 per cent) and the rotor speed was 60 000 r.p.m. The following boundary position vs. time data were obtained for an 0·4 per cent enzyme solution. (The time of the first photograph has been taken as zero.)

Number of photograph	Time (s)	r (cm)	$\ln r$
1	0	6·2215	1·8280
2	360	6·2680	1·8355
3	720	6·3140	1·8428
4	1080	6·3610	1·8502
5	1440	6·4085	1·8576
6	1800	6·4560	1·8650

Determine the $s_{20,w}^o$ sedimentation coefficient of the enzyme.

The partial specific volume at 293 K is $0·729 \text{ cm}^3 \text{ g}^{-1}$; the temperature increment of $\bar{v}$ is $6 \times 10^{-4} \text{ cm}^3 \text{ g}^{-1} \text{ K}^{-1}$; the density of the buffer at 278 K and 293 K is $1·0000 \text{ kg dm}^{-3}$ and $0·9982 \text{ kg dm}^{-3}$ respectively, while the viscosity of the buffer at the corresponding temperatures is $1·5188 \times 10^{-3}$ and $1·0087 \times 10^{-3} \text{ kg m}^{-1} \text{ s}^{-1}$.

Solution

(a) *Calculation of apparent sedimentation coefficient*

The apparent sedimentation coefficient can be calculated from the slope of the $\ln r$ vs. t plot

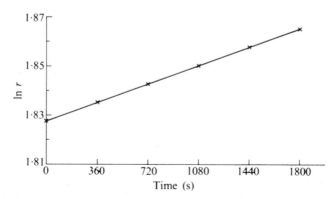

FIG. 6.5. Data for worked example 2.

From eqn (6.8)

$$S_{app} = \frac{d \ln r}{\omega^2 \, dt} = \frac{\text{slope}}{\omega^2} = \frac{2 \cdot 05 \times 10^{-5}}{\omega^2}$$

$$\omega^2 = \left(\frac{2\pi \, \text{r.p.m.}}{60}\right)^2 = \left(\frac{2\pi \times 6 \times 10^4}{60}\right)^2 = 3 \cdot 94 \times 10^7 \, s^{-2}$$

$$S_{app} = \frac{2 \cdot 05 \times 10^{-5}}{3 \cdot 94 \times 10^7} = 5 \cdot 2 \times 10^{-13} \, s$$

$S_{app} = 5 \cdot 2$ S (Svedberg).

(b) *Correction to standard conditions*
   This sedimentation coefficient has to be corrected to standard conditions (293 K in water).

The correction takes into account the variations of η, $\bar{v}$ and ρ with temperature. A further correction will still be needed to give the value at infinite dilution. To remember that this value of s is for a 0·4 per cent enzyme solution we shall use the notation $s_{app}^{0.4\%}$. (By convention for any concentration c, the term is written s_{app}^c etc.) If the actual measurements are performed in a buffer solution (b) at temperature T, then we can calculate an expression† for the value in water (w) at 293 K as

$$s_{20,w}^c = s_{app(T,b)}^c \cdot \frac{\eta_{T,b} \, (1 - \bar{v}\rho)_{293,w}}{\eta_{293,w}(1 - \bar{v}\rho)_{T,b}}$$

where the subscript T, b represents the value of the parameter in buffer at T. From the data given, we can calculate $\bar{v}_{278,b} = 0 \cdot 720$

$$s_{20,w}^{0.4\%} = 5 \cdot 2 \times \frac{(1 \cdot 5188 \times 10^{-3}) \times (1 - 0 \cdot 729 \times 0 \cdot 9982)}{(1 \cdot 0087 \times 10^{-3}) \times (1 - 0 \cdot 720 \times 1 \cdot 0000)} \, S$$

$$= 7 \cdot 565 \, S$$

Note the *large difference* between the real and approximate values and *the necessity for the correction.*

To calculate the $s_{20,w}^o$ value one further correction is required as illustrated in the following worked example.

(3) In the above example on D-glyceraldehyde-3-phosphate dehydrogenase the following $s_{20,w}^c$ values were also obtained at 0·2, 0·3, and 0·5 per cent protein concentrations:

$$s_{20,w}^{0.2\%} = 7 \cdot 72S; \qquad s_{20,w}^{0.3\%} = 7 \cdot 65S; \qquad s_{20,w}^{0.5\%} = 7 \cdot 48S.$$

Calculate the $s_{20,w}^o$ value.

---

† Note that $s = M(1 - \bar{v}\rho)/f$—then s depends on the temperature through $\bar{v}$, ρ, and η (since f is assumed to be directly proportional to η).

Solution

Plot these values of sedimentation coefficient vs. concentration and extrapolate to zero concentration (Fig. 6.6). This gives $s^{\circ}_{20,\text{w}} = 7\cdot86$ S.

This corrected and extrapolated $s^{\circ}_{20,\text{w}}$ can now be used as a new molecular parameter of the dissolved macromolecule. Its value is related to the molecular weight, shape and density of the particle.

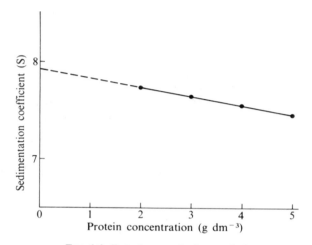

FIG. 6.6. Data from worked example 3.

From the shape of the boundary during the sedimentation run, additional qualitative information can be obtained. A single symmetrical boundary reflects a homogeneous sample. A sharp, slowly spreading boundary is characteristic of particles with loose extended structures.

It is possible to determine the molecular weight from the sedimentation and diffusion coefficients (or viscosities of the solution). However, obtaining the accurate corrected values of these parameters involves lengthy and tedious procedures and numerous conditions must be fulfilled (homogeneity, ideality, etc.) as we have seen. Most of these difficulties can be avoided by using sedimentation equilibrium.

Sedimentation equilibrium

If the rotor speed is not great enough to pile up all the material at the bottom of the cell, after a certain time a concentration gradient will be formed by the sedimentation transport. This gradient is large enough to drive a *diffusion transport which balances the sedimentation transport* in all parts of the cell. This state is the sedimentation equilibrium. *Because this is an equilibrium state, there is no dependence on the shape of the solute or the viscosity of the solution.* An exact thermodynamic treatment of sedimentation equilibrium is therefore possible (this type of treatment is not possible in the case of sedimentation velocity where the system is not at equilibrium).

By considering the distribution of molecules in an external gravitational field, a simple equation can be derived to describe the equilibrium concentration gradient for a single solute component in an ideal two component solution

$$\frac{1}{c}\frac{dc}{dr} = \frac{\omega^2 r M (1 - \bar{v}\rho)}{RT} \tag{6.10}$$

where c, the concentration, is a function of r, the radial position, but not a function of time (since the system is at equilibrium). This equation relates the term $M(1 - \bar{v}\rho)$, the effective molecular weight after correction for buoyancy, to the equilibrium concentration distribution in a cell spinning with angular velocity ω.

The validity of eqn (6.10) can be extended to solutions containing more than one solute component. Eqn (6.10) treated in various ways will yield the number, weight or z-average molecular weight. The most commonly used equation yields the weight-average molecular weight $\bar{M}_w$

$$\boxed{\bar{M}_w = \frac{2RT}{(1 - \bar{v}\rho)\omega^2}\frac{d\ln c}{dr^2}} \tag{6.11}$$

To calculate the molecular weight the concentration as a function of radial distance has to be known. This can be directly obtained from equilibrium runs monitored by a photoelectric scanner. However a more widely used method in sedimentation equilibrium because of its precision and sensitivity is the Rayleigh interferometer optics system. There is a disadvantage in this method in that the fringe displacements are proportional not to the concentration, but to the concentration difference between the meniscus and the point of interest, $(c_r - c_m)$. This difficulty can be eliminated in two ways.

(i) Using the meniscus depletion method. In this method a high rotor speed is applied to reduce the solute concentration at the meniscus to zero $c_m = 0$. In this way the fringe displacement throughout the cell (ΔI) will be proportional to the concentration c. The molecular weight can be calculated by plotting $\ln \Delta I$ vs. r^2, and using the slope† of this plot in place of $d\ln c/dr^2$ in eqn (6.11) as in the worked example on p. 83.

---

† Since $\Delta I = k \times c$, then $\ln I = \ln k + \ln c$, i.e. the plot of $\ln \Delta I$ vs. r_i^2 will have the same *slope* as that of $\ln c$ vs. r_i^2.

(ii) By rewriting eqn (6.10) and extending to a multicomponent system, we can obtain $\bar{M}_w$ as:

$$\bar{M}_w = \frac{2RT}{(1 - \bar{v}\rho)\omega^2} \frac{c_r - c_m}{c_0(r_{\,}^2 - r_m^2)}$$

where c_0 is the initial concentration of the solute. $c_r - c_m$ can be directly obtained in ΔI fringe displacement units from the equilibrium run. c_0 can be determined (also in ΔI fringe displacements) from an additional synthetic boundary run, (where solvent is layered over the solution in an appropriate cell, thus making the junction between the liquids the boundary).

In the case of polydisperse or reversibly associating systems the molecular weight distribution of the sample will depend on the radial position in the cell.† In such cases the actual value of the n-, w-, and z-average molecular weights *at each point* of the cell can be calculated, and the molecular weight of the components, their relative abundance, and the dissociation constant can be estimated.

The following equations can be used for calculations of the molecular weights in the meniscus depletion method.

$$\bar{M}_n = \frac{M_1 c}{\displaystyle\int_0^c \frac{M_1}{M_w(c)} \, dc}$$

$$\bar{M}_z = \frac{2RT}{(1 - \bar{v}\rho)\omega^2} \frac{d^2c/d(r^2)^2}{dc/d(r^2)}$$

$\bar{M}_w$ can be calculated from eqn (6.11). M_1 is the molecular weight of the monomer in the case of reversibly associating systems.

The analysis of interacting systems is complicated. However, a monomer–dimer–tetramer equilibrium can be described in terms of dissociation constants by the analysis of the concentration distribution at sedimentation equilibrium.

To summarize, the sedimentation equilibrium method is a means of avoiding most of the difficulties encountered with the sedimentation velocity method. Using interferometer optics the experiments can be carried out at low concentrations (0·01–0·03 per cent) such that no correction for concentration dependence is necessary. Using a short solution column (0·15–0·30 cm) the time requirement for the experiment (i.e. to reach equilibrium) can be drastically reduced to 4–8 hours. An added advantage of the meniscus depletion method is that the molecular weight of the sample

---

† A steeper concentration gradient will be established for the larger than for the smaller species in polydisperse systems. Thus more associated molecules will be present near the bottom of the cell than near the meniscus.

can be determined even in the presence of small amounts of large aggregates as shown in the worked example below.

A single meniscus depletion sedimentation equilibrium experiment is sufficient to determine the molecular weight of a macromolecule and to judge the homogeneity of the sample. The experiment requires 20–30 μg of material and can be carried out in 5–6 hours.

Note on the measurement of $\bar{v}$

The partial specific volume of most proteins is about $0\cdot7$ cm^3 g^{-1}. The Svedberg equation contains the term $(1 - \bar{v}\rho)$ and so an error of 1 per cent in $\bar{v}$ (in water at 293 K) will result in a 3 per cent error in M. This is in contrast to an error of 1 per cent in either s or D which result in a 1 per cent error in M—because of the linear dependence. This illustrates the requirement to know $\bar{v}$ much more precisely than s or D. The most accurate method of measuring $\bar{v}$ uses a pycnometer—but this requires large amounts of sample. $\bar{v}$ can also be estimated for proteins in a particular solvent from the chemical composition of the solution. This estimation is based on the partial specific volume of each type of amino acid present. The most accurate method of calculating $\bar{v}$ is from density measurements, and by use of the mechanical oscillator densimeter a precision of 10^{-6} kg dm^{-3} can be achieved.

The use of two solvents of different densities in the analytical ultracentrifuge avoids the requirement for an independent determination of $\bar{v}$. For instance one ultracentrifuge run in H_2O and one in a mixture of H_2O/D_2O leads to two simultaneous equations which can be solved for $\bar{v}$ and M. This is illustrated in the worked example.

Worked example

Two meniscus depletion sedimentation equilibrium experiments were performed with an immunoglobulin sample. In the first run H_2O was used as solvent and in the second a 20 per cent H_2O–80 per cent D_2O mixture.

The initial protein concentration was $0\cdot03$ per cent and a 16 000 r.p.m. rotor speed was selected†; both runs were conducted at 298 K.

Rayleigh interferometer optics were used to take photographs after reaching the sedimentation equilibrium. The scheme of such a photograph is shown in Fig. 6.7

The fringe displacements ΔI_i were measured as a function of radial distance r_i. The data are given in Tables 6.2 and 6.3 and represent typical experimental measurements.

---

† If the approximate molecular weight is known the optimal rotor speed (r.p.m.) can be selected according to the following relationship: $6 \times 10^6/\sqrt{M}$ for the meniscus depletion method and $2\cdot5 \times 10^6/\sqrt{M}$ for the conventional (low speed) sedimentation equilibrium method—see Problem 9.

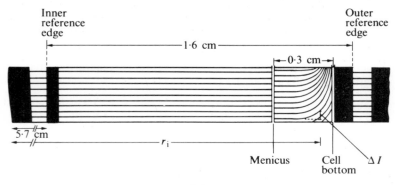

(a)

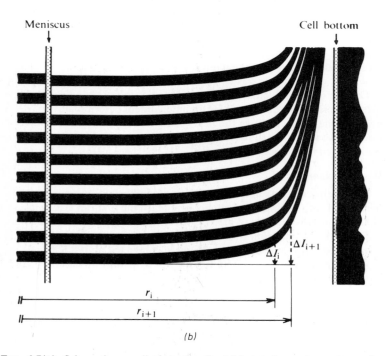

(b)

FIG. 6.7(a). Schematic overall view of a Rayleigh interferometer pattern of a meniscus depletion sedimentation equilibrium concentration distribution and (b) enlargement of the section in (a) showing the fringe displacements reflecting the equilibrium concentration distribution.

The densities of H_2O and D_2O at 298 K are 0·997 and 1·105 kg dm^{-3} respectively $(R = 8·31 \text{ J K}^{-1} \text{ mol}^{-1})$. Calculate (a) the partial specific volume of the immunoglobulin and (b) its molecular weight, and (c) comment on the homogeneity of the sample, and on the necessity of corrections.

Solution

Calculate the r^2 and the corresponding $\ln \Delta I_i$ values. These are given in Tables 6.2 and 6.3. Then plot $\ln \Delta I$ vs. r_i^2 (Figs. 6.8 and 6.9).

From the graphs we can determine the slope of the linear part of the plots. (The increase in the slope near the cell bottom reflects the presence of higher molecular weight components, possibly aggregates). We then use eqn (6.11) for the calculation as follows:

$$M = \frac{2RT}{(1-\bar{v}\rho)\omega^2} \frac{\mathrm{d}\ln c}{\mathrm{d}\,r^2} = \frac{2RT}{(1-\bar{v}\rho)\omega^2} \underbrace{\frac{\ln \Delta I}{r^2}}_{\text{slope}}$$

TABLE 6.2
Data for worked example (experiment in H_2O)

No.	r_i(cm)	r_i^2 (cm^2)	ΔI_i (μm)	$\ln \Delta I_i$
r_m	6·7054		Effectively −2	
2	6·7100		no 0	
3	6·7250		change, 3 $\Delta I_i = 0$	
4	6·7400		just −1	
5	6·7550		random 0	
6	6·7700		scatter	
7	6·7850			
8	6·8000			
9	6·8150			
10	6·8300	46·649	26	3·266
11	6·8450	46·854	72	3·749
12	6·8600	47·060	69	4·233
13	6·8750	47·266	122	4·718
14	6·8900	47·472	182	5·204
15	6·9050	47·679	296	5·691
16	6·9200	47·886	483	6·179
17	6·9350	48·094	788	6·669
18	6·9500	48·303	1733	7·458
19	6·9650	48·511	3837	8·253
20	6·9800	48·720	9740	9·184
21	6·9950	48·930	— †	—
r_b	7·0250	49·351	— †	—

† Difficult to make precise measurements near to the bottom of the cell.

TABLE 6.3
Data for worked example (experiment in H_2O/D_2O)

No.	r_i (cm)	r^2 (cm^2)	ΔI_i (μm)	ln ΔI_i
r_m	6·7125		Effectively 0	
2	6·7250		no change 5	
3	6·7400		just random −4 $\Delta I_i = 0$	
4	6·7550		scatter 0	
5	6·7700			
6	6·7850			
7	6·8000			
8	6·8150			
9	6·8300	46·649	24	3·165
10	6·8450	46·854	35	3·540
11	6·8600	47·060	50	3·916
12	6·8750	47·266	73	4·293
13	6·8900	47·472	107	4·671
14	6·9050	47·679	156	5·050
15	6·9200	47·886	228	5·430
16	6·9350	48·094	334	5·811
17	6·9500	48·303	687	6·532
18	6·9650	48·511	1468	7·292
19	6·9800	48·720	3133	8·050
20	6·9950	48·930	—†	—
r_b	7·0074	49·104	—†	—

† Difficult to make precise measurements near the bottom of the cell.

from which by rearranging

$$M(1 - \bar{v}\rho) = \frac{2RT}{\omega^2} \cdot \text{slope}.$$

We have two parameters M and $\bar{v}$ to determine and two equations from the two runs:

$$M(1 - \bar{v}\rho_1) = \frac{2RT}{\omega^2} (\text{slope } 1) = A$$

$$M(1 - \bar{v}\rho_2) = \frac{2RT}{\omega^2} (\text{slope}) 2 = B.$$

Now A and B can be calculated from the experimental results.

The above equations can be rearranged to make $\bar{v}$ and M the subjects. Thus

$$\bar{v} = \frac{A - B}{A\rho_2 - B\rho_1} \quad \text{and} \quad M = \frac{A}{1 - \bar{v}\rho_1}$$

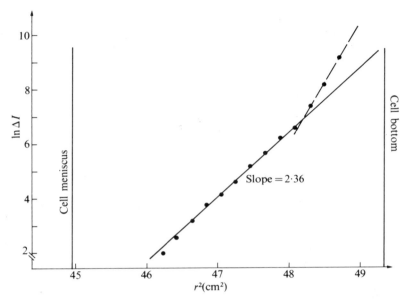

FIG. 6.8. Plot of data from Table 6.2 of worked example.

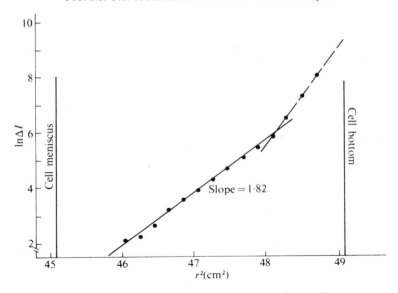

FIG. 6.9. Plot of data from Table 6.3 of worked example.

so from the values of A and B, $\bar{v}$ and M can be calculated. The density of the solvent in the second experiment is:

$$\rho_2 = 0 \cdot 2 \rho_{H_2O} + 0 \cdot 8 \rho_{D_2O} = (0 \cdot 2 \times 0 \cdot 997) + (0 \cdot 8 \times 1 \cdot 105)$$

$$= 1 \cdot 083 \text{ kg dm}^{-3}.$$

Noting that $\omega^2 = (2\pi \text{ r.p.m.}/60)^2 = 2 \cdot 8 \times 10^6 \text{ s}^{-2}$ and that the slopes in Figs. 6.8 and 6.9 have to be multiplied by 10^4 to bring them into units of m^{-2}, we have

$$A = \frac{2 \times 8 \cdot 31 \times 298 \times 2 \cdot 36 \times 10^4}{2 \cdot 8 \times 10^6} = 41 \cdot 8 \text{ kg mol}^{-1}$$

and

$$B = \frac{2 \times 8 \cdot 31 \times 298 \times 1 \cdot 82 \times 10^4}{2 \cdot 8 \times 10^6} = 32 \cdot 2 \text{ kg mol}^{-1}$$

so that

$$A\rho_2 = 41 \cdot 8 \times 1 \cdot 083 = 45 \cdot 3$$

and

$$B\rho_1 = 32 \cdot 2 \times 0 \cdot 997 = 32 \cdot 1$$

$$\therefore \quad \bar{y} = \frac{41 \cdot 8 - 32 \cdot 2}{45 \cdot 3 - 32 \cdot 1} = 0 \cdot 727 \text{ dm}^3 \text{ kg}^{-1} \text{ (or cm}^3 \text{ g}^{-1})$$

(a) The partial specific volume of immunoglobulin in water at 298 K is

$$0 \cdot 727 \text{ cm}^3 \text{ g}^{-1}.$$

(b) Now $(1 - \bar{v}\rho_1) = (1 - 0 \cdot 727 \times 0 \cdot 997) = 0 \cdot 275$, hence

$$M = \frac{41 \cdot 8}{0 \cdot 275} = 152 \text{ kg mol}^{-1}$$

The molecular weight of this immunoglobulin is 152 000.

(c) The sample contains some high molecular weight contamination as can be seen in the upwards curvature of the $\ln c (\Delta I)$ vs. r^2 plot. This contamination may be aggregated protein, which is accumulated near the cell bottom in the spinning cell.

Since the protein concentration is very low in the experiment no extrapolation to zero concentration is necessary.

The temperature dependence is taken into account in the equation used in the calculations, since the equations contain a term in T.

Approach to equilibrium

If sedimentation equilibrium is reached, the net flow of material is zero in all parts of the cell. As has been shown previously it is very convenient to calculate molecular weight from the equilibrium concentration distribution of a solute in the spinning ultracentrifuge cell.

It takes several hours (depending on the length of the solution column in the cell) to reach equilibrium. However, there are two points in the cell, the meniscus and the bottom, where obviously there is no transport of matter in any phase of the experiment. From the concentration distribution function near to those surfaces the molecular weight can be calculated, before the total equilibrium is reached if the angular velocity ω is known. This method is named after Archibald, who first pointed out the possibility of its application.

The advantage of the method is the short time required for the experiment. The disadvantage is its inferiority in accuracy (because the precision of the optical method is poorest near the extremities) and the computation of the results is complicated and time consuming.

For polydisperse systems the method will result in low molecular weight species collecting at the meniscus and higher molecular weight species collecting at the bottom of the cell. (This follows directly from the molecular weight dependent transport of matter by sedimentation.) By extrapolation to zero time, the weight average molecular weight can be obtained. The Archibald method has been used less frequently since short column equilibrium experiments became popular. These allow *more* precise results from runs of reasonable duration.

PROBLEMS

1. A solution containing $0 \cdot 5$ g ribonuclease in $0 \cdot 1$ dm^3 of $0 \cdot 2$ mol dm^{-3} sodium chloride exerted an osmotic pressure of $0 \cdot 0097$ atm at 298 K. The membrane is permeable to all species except ribonuclease. Determine the molecular weight of ribonuclease. How and why (qualitatively) would the result have been different if the experiment had been carried out in the presence of a much smaller amount of sodium chloride? ($R = 0 \cdot 082$ dm^3 atm mol^{-1} K^{-1}.)

2. Calculate the difference in osmotic pressure between capillary blood and the tissue fluid at 310 K. Assume that this difference arises because of the protein in the blood (concentration = 7 per cent; molecular weight 66 000), which is absent in the tissue fluid. (Ignore the Donnan effect.) ($R = 0 \cdot 082$ dm^3 atm mol^{-1} K^{-1}.)

3. The osmotic pressure of blood plasma is $7 \cdot 6$ atm at 310 K. What is the total concentration of dissolved species present? Use this result to calculate the freezing point depression of the plasma. What would the freezing point depression be if the plasma were dialysed to remove small solutes? The protein concentration and molecular weight can be assumed to be the same as in Problem 2.

Isotonic saline is that concentration of NaCl which prevents osmotic disruption of the blood cells. Comment on the fact that isotonic saline is 0·95 per cent NaCl (by weight).

4. Calculate the difference in osmotic pressure between the blood plasma of a diabetic patient and that of a healthy individual if the difference is assumed to be solely due to the higher level of glucose in the plasma of the diabetic patient. (The plasma glucose levels are 1·80 and 0·85 g dm^{-3} respectively, $T = 310$ K.)

5. The antifreeze protein from an Antarctic fish was isolated and found to have a molecular weight of 17 000. What should the freezing point depression of the serum be if the protein solution behaves ideally? (It can be assumed that the protein concentration is 10 g dm^{-3}, and that the serum has been dialysed to remove small molecules.) Compare your result with the observed freezing point depression of 0·6 K and with the freezing point depression of a 10 g dm^{-3} solution of lysozyme (molecular weight 14 500) which is 0·0014 K.

6. Consider a protein of molecular weight 60 000, which carries six negative charges at pH 7·4, placed in a cell enclosed by a membrane permeable to all ions except the cell protein. Assume the protein (at a concentration of 60 g dm^{-3}) is added as its sodium salt and 0·15 mol dm^{-3} NaCl is also added to the protein side. The volume of water on the other side of the membrane is the same as the volume of solution on the protein side. Calculate the equilibrium concentrations of ions on both sides of the membrane.

   This situation approximates to that in the kidney. The observed concentrations of Na$^+$ and Cl$^-$ ion on the protein (plasma) side are 0·120 mol dm^{-3} and those on the filtrate (water) side are 0·6 mmol dm^{-3}. Comment.

7. The concentrations of Na$^+$ and K$^+$ ions in tissues are about 11 and 92 mmol dm^{-3} respectively (inside) and 140 and 4 mmol dm^{-3} respectively (outside). Calculate the free energy requirement for maintenance of each of these ion gradients. ($T = 310$ K.)

8. In a study of the sedimentation of pancreatic α-amylase at 298 K and a rotor speed of 50 000 r.p.m. the following set of data was obtained.

t(s)	0	360	720	1080	1440	1800
boundary (cm)	6·236	6·267	6·298	6·330	6·362	6·393

   (a) Noting that the ratio of viscosities of the solvent at 298 and 293 K is 0·8872 calculate the $s_{20,w}$ sedimentation coefficient.
   (b) Use this value to calculate the molecular weight of α-amylase. ($D_{293,w} = 7·62 \times 10^{-11}$ m^2 s^{-1}; $\bar{v} = 0·728$ cm^3 g^{-1}; $R = 8·31$ J K^{-1} mol^{-1}.) (Assume the solution density is 1·000 kg dm^{-3}.)

9. The equilibrium concentration distribution in a sedimentation equilibrium experiment for a given system at a given temperature is determined by the rotor speed. In a conventional sedimentation experiment the optimal ratio of the concentration at the meniscus to that at the cell bottom is 1:4 for precise calculation. In the case of a meniscus depletion experiment the concentration at the meniscus should be almost zero.
   (a) Calculate the optimal rotor speed for a conventional (low speed) sedimentation equilibrium experiment for a protein sample of molecular weight 100 000 at 293 K if the position of the meniscus r_m and the bottom of the cell r_b are 6·7

and 7·0 cm respectively from the axis of rotation. ($\bar{v} = 0\cdot75$ cm^3 g^{-1}; density of the solution $1\cdot00$ kg dm^{-3}.)

(b) Calculate the optimal rotor speed for a meniscus depletion experiment for the same system. (A concentration ratio at the meniscus to that at the bottom of the cell $c_m : c_b$ of $1:1000$ is required.)

(c) Make logarithmic plots for the required r.p.m. values vs. molecular weight for both types of equilibrium.

10. A meniscus depletion (high speed) sedimentation equilibrium experiment was carried out at 277 K with a ceruloplasmin sample of 0·02 per cent initial concentration. Rayleigh interferometer optics were used and the rotor speed was 16 000 r.p.m. To reduce the time required to reach equilibrium a 0·3 cm solution column was used in a double sector cell. The following set of data was obtained by measuring the fringe displacement as a function of radial distance in the interference pattern. (A schematic diagram of the interference pattern and the way of analysing it is shown in Fig. 6.7(b))

No.	r_i (cm)	ΔI_i (μm)	No.	r_i (cm)	ΔI_i (μm)
r_m	6·7083	0	12	6·8650	65
2	6·7150	0	13	6·8800	107
3	6·7300	0	14	6·8950	194
4	6·7450	0	15	6·9100	347
5	6·7600	0	16	6·9250	631
6	6·7750	0	17	6·9400	1141
7	6·7900	−5	18	6·9550	2059
8	6·8050	5	19	6·9700	4230
9	6·8200	10	20	6·9850	12330
10	6·8350	19	21	7·0000	38170
11	6·8500	34	22	7·0150	—
			r_b	7·0270	—

(a) Calculate the molecular weight of cerulopla

(b) Comment on the homogeneity of the sample.

($\bar{v}$ of ceruloplasmin at 277 K is 0·713 cm^3 g^{-1} and the density of the solution was $1\cdot00$ kg dm^{-3} ($R = 8\cdot31$ J K^{-1} mol^{-1}).)

11. One component of the serum lipoprotein has a molecular weight of $2\cdot3 \times 10^6$. It is suspended in 1·5 mol dm^{-3} NaCl. In a sedimentation velocity experiment, the boundary forms at the cell bottom and migrates upwards to the meniscus.

(a) Comment on this, and (b) Suggest a method to separate lipoprotein fractions of molecular weights of 180 000 and $2\cdot3 \times 10^6$ with different protein–lipid composition.

12. Proteins treated with sodium dodecyl sulphate (SDS) detergent unfold and the subunits (if present) dissociate. A native protein in a buffer solution has a sedimentation coefficient of 7·6 S. After SDS treatment the protein has a sedimentation coefficient of 2·6 S.

Can you decide by ultracentrifugation whether unfolding or unfolding and dissociation take place? If not suggest the appropriate experiment.

13. A number of proteins were studied by gel electrophoresis on 7·5 per cent acrylamide gels (in the presence of sodium dodecylsulphate) and by gel filtration (using a column of Sephadex G-200). The following data were obtained:

Protein	Molecular weight	Number of sub-units	Relative mobility (electro-phoresis)	Elution volume (gel filtration) (cm^3)
Phosphorylase b	200 000	2	0·25	105
Fumarase	198 000	4	0·53	106
Aldolase	160 000	4	0·64	114
Hexokinase	102 000	4	0·83	141
Bovine serum albumin	66 000	1	0·43	168
Ovalbumin	45 000	1	0·58	190
Carbonic anhydrase	29 000	1	0·77	216
Trypsinogen	24 000	1	0·87	230
Myoglobin	17 000	1	1·00	242

Under these conditions, the relative mobilities of creatine kinase and glyceralde-hyde-3-phosphate dehydrogenase on electrophoresis were 0·63 and 0·67 respectively, and their elution volumes on gel filtration were 154 and 122 cm^3 respectively. Comment on these data and evaluate the probable subunit composition of these enzymes.

7. Acids and bases

Introduction

THIS chapter deals with another important application of our thermodynamic principles, namely the proton equilibria in aqueous solutions.

Water dissociates according to the equation†

$$H_2O \rightleftharpoons H^+ + OH^-$$

For this process we can write a dissociation constant, K

$$K = \frac{a_{H^+} \cdot a_{OH^-}}{a_{H_2O}}.$$

Since a_{H_2O} (based on a standard state of water as a 1 mol dm^{-3} solution) is so large as to be effectively constant, we can define a new constant, K_w, by multiplying K by a_{H_2O}:

$$K_w = a_{H^+} \cdot a_{OH^-}.$$

For dilute solutions, we can ignore the correction due to non-ideality and set the activities of the ions equal to their concentrations. (This is equivalent to setting the activity coefficients for these ions equal to units.)

$$\therefore \quad \boxed{K_w = [H^+][OH^-]} \quad .$$

K_w is called the *ionic product for water*. It has a value 10^{-14} at 298 K and increases with temperature (see Problem 1).

This means that in any aqueous solution at 298 K, the product of the concentrations of H^+ and OH^- is 10^{-14}. In a neutral solution, $[H^+] = [OH^-] = 10^{-7}$ (mol dm^{-3}).

The concept of pH

The pH of a solution is a convenient way of expressing the concentration of hydrogen ions (H^+) in solution, to avoid the use of large negative powers of 10 (as in the previous paragraph).

---

† In aqueous solutions the proton exists in a hydrated form (H_3O^+). In this chapter we shall use H^+ as an abbreviation for H_3O^+ and remember tnat all species are in their appropriate hydrated forms.

The pH of a solution is defined as

$$pH = -\log_{10}[H^+]$$.

Thus the pH of a neutral solution (where $[H^+] = 10^{-7}$) is 7. In fact the usual pH range for most experiments of interest is from 0 (corresponding to a 1 mol dm^{-3} solution of H^+ ions) to 14 (corresponding to a 1 mol dm^{-3} solution of OH^- ions).

Acids and bases

The most useful definition of acids and bases is that due to Brønsted and Lowry.† Acids are defined as potential proton donors and bases as potential proton acceptors. Each acid is therefore related to its *conjugate base* (and vice versa) by an equilibrium of the type:

$$CH_3CO_2H + OH^- \rightleftharpoons CH_3CO_2^- + H_2O$$

| Acid I | Conjugate Base II | Conjugate Base I | Acid II |

The strength of an acid is thus measured by its ability to provide protons. Hydrochloric and nitric acids, for example, are almost completely dissociated in aqueous solutions and are therefore termed *strong acids*. By contrast acetic and formic acids are only slightly dissociated in solution and termed *weak acids*.

In *aqueous* solutions the proton acceptor is H_2O and hence (H_3O^+) is the strongest acid. The strengths of other acids are then measured by the degree to which they produce H_3O^+ ions, by dissociation. For example, consider an acid HA dissociating according to

$$HA + H_2O \rightleftharpoons H_3O^+ + A^-.$$

The *equilibrium constant* (K) for this reaction is defined as

$$K = \frac{a_{H_3O^+} \cdot a_{A^-}}{a_{HA} \cdot a_{H_2O}},$$

By convention, this constant is multiplied by a_{H_2O} to give the *acid dissociation constant* for aqueous solution (K_a)

$$\therefore \quad K_a = \frac{a_{H_3O^+} \cdot a_{A^-}}{a_{HA}}.$$

---

† A more general definition of acids and bases was given by Lewis. Acids tend to accept electrons and bases tend to donate electrons. In this definition, for instance BCl_3 is an acid since it will accept electrons from $(C_2H_5)_3N$, which is thus a base.

Setting the activities equal to the concentrations and using H^+ as the abbreviation for H_3O^+, we obtain

$$K_a = \frac{[H^+][A^-]}{[HA]}.\dagger$$

The larger K_a, the greater the degree of dissociation of HA and hence the stronger the acid.

Similarly for the conjugate base (A^-) we can write

$$A^- + H_2O \rightleftharpoons HA + OH^-$$

and the *dissociation constant of the base* (K_b) is defined by considering the above equilibrium:

$$K_b = \frac{[HA][OH^-]}{[A^-]},$$

Combining the above definitions we obtain the relationship

$$K_a \cdot K_b = \frac{[H^+][A^-]}{[HA]} \times \frac{[HA][OH^-]}{[A^-]}$$

$$= [H^+][OH^-],$$

i.e.

$$\boxed{K_a \cdot K_b = K_w},$$

where K_w is the ionic product of water. Thus the product of the acid and base dissociation constants of a compound is 10^{-14} at 298 K.

In the same way that pH is used to denote the concentration of $[H^+]$, the symbols pK_a and pK_b are used for $-\log_{10} K_a$ and $-\log_{10} K_b$ respectively. Clearly the lower the value of pK_a of an acid, the stronger the acid. Now since

$$K_a \cdot K_b = K_w,$$

we obtain by taking logarithms

$$\boxed{pK_a + pK_b = 14}.$$

Worked examples

1. The pK_a of formic acid is 3·77 (at 298 K). What is the pH of a 0·01 mol dm^{-3} solution of formic acid in water?

---

† The value of K_a is applicable to dilute aqueous solutions. In concentrated solutions, activities will be different from concentrations and the term a_{H_2O} is no longer the same. It should also be pointed out that the value of K must be corrected for the a_{H_2O} term if the behaviour of the acid is being studied in some other solvent system (e.g. water/acetone).

Solution
$$pK_a = -\log_{10} K_a,$$

$$\therefore \quad K_a \text{ for formic acid} = 1\cdot695 \times 10^{-4} \text{ (mol dm}^{-3}).$$

Consider the dissociation of HCO_2H as:

$$HCO_2H \rightleftharpoons H^+ + HCO_2^-.$$
$$(0\cdot01 - x) \qquad x \qquad x$$
$$\text{mol dm}^{-3} \quad \text{mol dm}^{-3} \quad \text{mol dm}^{-3}$$

Let the concentration of $H^+ = x$ mol dm^{-3}. Then the concentration of HCO_2^- is also x mol dm^{-3} and the concentration of undissociated HCO_2H is $(0\cdot01 - x)$ mol dm^{-3}. Now

$$K_a = \frac{[H^+][HCO_2^-]}{[HCO_2H]}$$

$$1\cdot695 \times 10^{-4} = \frac{x^2}{(0\cdot01 - x)},$$

$$x^2 + (1\cdot695 \times 10^{-4})x - (1\cdot695 \times 10^{-6}) = 0.$$

From which $x = 1\cdot22 \times 10^{-3}$ (taking the positive root).

Thus the concentration of H^+ is $1\cdot22 \times 10^{-3}$ mol dm^{-3}, and from the formula $pH = -\log_{10}[H^+]$, we calculate that the pH of the solution is $2\cdot91$. We have assumed here that all the protons in the solution are contributed by the dissociation of HCO_2H. For very dilute solutions of acids or for very weak acids we would have to take account of the contribution of protons from dissociation of the solvent water.

2. The pK_b of methylamine is $3\cdot4$ at 298 K. What is the pH of a $0\cdot01$ mol dm^{-3} solution?

Solution
$$pK_b = -\log_{10} K_b,$$

$$\therefore \quad K_b \text{ for methylamine} = 3\cdot975 \times 10^{-4} \text{ (mol dm}^{-3}).$$

For the equilibrium

$$CH_3NH_2 + H_2O \rightleftharpoons CH_3NH_3^+ + OH^-$$
$$(0\cdot01 - x) \qquad\qquad x \qquad\qquad x$$
$$\text{mol dm}^{-3} \qquad\qquad \text{mol dm}^{-3} \quad \text{mol dm}^{-3}$$

let the concentration of $OH^- = x$ mol dm^{-3}, ($= [CH_3NH_3^+]$). The concentration of CH_3NH_2 is $(0\cdot01 - x)$ mol dm^{-3}. Now

$$K_b = \frac{[CH_3NH_3^+][OH^-]}{[CH_3NH_2]},$$

$$\therefore \quad 3\cdot975 \times 10^{-4} = \frac{x^2}{(0\cdot01 - x)},$$

$$x^2 + (3\cdot975 \times 10^{-4})x - (3\cdot975 \times 10^{-6}) = 0.$$

From which $x = 1\cdot8 \times 10^{-3}$ (taking the positive root).

Thus the concentration of OH^- is 1.8×10^{-3} mol dm^{-3} and since $[H^+][OH^-] = 10^{-14}$, we calculate that $[H^+] = 5.55 \times 10^{-12}$ mol dm^{-3}, i.e. the pH of the solution is $\underline{11.26}$.

Buffer solutions

When a strong base (e.g. NaOH) is added to a strong acid (e.g. HCl), neutralization is virtually complete after the addition of an equivalent amount of base. (The dissociation of H_2O is so slight that its effect can be ignored.) A graph of pH against equivalents of base added shows that the pH changes very rapidly around the equivalence point.†

The situation is somewhat different when a weak acid is neutralized by a strong base (the dotted curve in Fig. 7.1). This different behaviour can be explained by considering the equilibrium for the weak acid, HA.

$$K_a = \frac{[H^+][A^-]}{[HA]}.$$

Taking logarithms and rearranging, we obtain

$$\boxed{pH = pK_a - \log_{10} \frac{[HA]}{[A^-]}.} \qquad (7.1)$$

This is often termed the *Henderson–Hasselbalch* equation.

In our example A^- is produced by neutralization of the weak acid. We can use the Henderson–Hasselbalch equation to calculate the pH of the solution as a function of the amount of strong base added (i.e. the amount of A^- produced). This would give the dotted curve shown in Fig. 7.1. We see that the slope of the curve is a minimum when 0.5 equivalents of base have been added (i.e. when $[A^-] = [HA]$).‡ Thus at this point addition of base causes the least change in pH. A solution which resists pH changes is known as a *buffer solution*. We can see that the buffer action results from the ability of A^- to take up added H^+ so as to form HA thereby reducing the pH change. Conversely addition of OH^- will take up protons from HA (leading to formation of H_2O and A^-), again reducing the pH change.

To summarize, the buffer capacity of a solution of a weak acid (HA) and its conjugate base (A^-) is a maximum when $[A^-] = [HA]$, i.e. when the pH of the solution is equal to the pK_a of the weak acid. Also the pH change

---

† An indicator is a weak acid whose ionization is accompanied by a change in colour. An indicator is chosen for a particular titration if its pK_a, (usually denoted by pK_{In}) is approximately the pH of the equivalence point.

‡ This can be seen from the Henderson–Hasselbalch equation directly, since the changes in the term $\log_{10}[HA]/[A^-]$ are smallest for a given change in $[HA]$ and $[A^-]$ when $[HA] = [A^-]$.

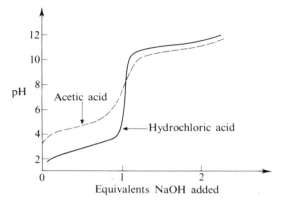

FIG. 7.1. Titration of hydrochloric acid (a strong acid) (solid line) and acetic acid (a weak acid) (broken line) with sodium hydroxide (a strong base).

resulting from addition of a given amount of acid or alkali will be smaller, the greater the concentration of [HA] and [A⁻], i.e. the more concentrated the buffer solution.

Weak bases and their conjugate acids can also be used as buffers and they will be most effective when the pH of the solution is around the pK_a (i.e. around $14 - pK_b$).

Worked example

(a) Tris is a weak base ($K_a = 8 \cdot 3 \times 10^{-9}$ (mol dm⁻³)) commonly used as a buffer in biochemical systems. What is the ratio of the concentrations of tris and its conjugate acid at pH 8·0?

Solution

Now

$$pH = pK_a - \log_{10} \frac{[\text{tris-H}^+]}{[\text{tris}]},$$

and

$$pK_a = -\log_{10} K_a,$$

$$= 8 \cdot 08,$$

$$\therefore \quad 8 \cdot 0 = 8 \cdot 08 - \log_{10} \frac{[\text{tris-H}^+]}{[\text{tris}]},$$

$$\log_{10} \frac{[\text{tris}]}{[\text{tris-H}^+]} = -0 \cdot 08,$$

$$\therefore \quad \text{the ratio } \frac{[\text{tris}]}{[\text{tris-H}^+]} = 0\cdot83.$$

(b) If the total concentration of tris in the above solution is 100 mmol dm^{-3}, what is the change in pH after addition of 5 mmol dm^{-3} H$^+$? If the total concentration of tris is reduced to 20 mmol dm^{-3}, what will the new pH change be?

Solution

If

$$[\text{tris}] + [\text{tris H}^+] = 0\cdot1 \text{ mol dm}^{-3},$$

and

$$\frac{[\text{tris}]}{[\text{tris-H}^+]} = 0\cdot83 \text{ (from above)},$$

then

$$[\text{tris}] = 0\cdot045 \text{ mol dm}^{-3},$$

and

$$[\text{tris-H}^+] = 0\cdot055 \text{ mol dm}^{-3}.$$

The effect of adding $0\cdot005$ mol dm^{-3} H$^+$ is to reduce [tris] and increase [tris H$^+$] by this amount, i.e. after the addition, [tris] = $0\cdot040$ mol dm^{-3} and [tris-H$^+$] = $0\cdot060$ mol dm^{-3}. So from eqn (7.1)

$$\text{pH} = \text{p}K_a - \log_{10}\frac{[\text{tris-H}^+]}{[\text{tris}]},$$

$$\text{pH} = 8\cdot08 - \log_{10}\frac{[0\cdot06]}{[0\cdot04]},$$

$$\underline{\text{pH} = 7\cdot9.}$$

i.e. the addition of 5 mmol dm^{-3} H$^+$ has lowered the pH by $\underline{0\cdot1 \text{ units}}$. In the second case,

$$[\text{tris}] + [\text{tris-H}^+] = 0\cdot02 \text{ mol dm}^{-3},$$

and

$$\frac{[\text{tris}]}{[\text{tris H}^+]} = 0\cdot83,$$

then

$$[\text{tris}] = 0\cdot009 \text{ mol dm}^{-3}$$

and

$$[\text{tris-H}^+] = 0\cdot011 \text{ mol dm}^{-3}.$$

After addition of H^+, [tris] is reduced to 0.004 mol dm^{-3}, and [tris-H^+] $=$ 0.016 mol dm^{-3}.

As before

$$pH = pK_a - \log_{10} \frac{[\text{tris-H}^+]}{[\text{tris}]},$$

$$= 8.08 - \log_{10} \frac{[0.016]}{[0.004]},$$

$$pH = \underline{7.48}.$$

In this case the pH has changed by $\underline{0.52 \text{ units}}$. This illustrates the point made in the text, that buffer solutions are more effective at higher concentrations.

If no buffer had been present, the final pH would have been $\underline{2.3}$ (i.e. $-\log_{10} [0.005]$).

The concept of a buffer solution is not limited to protonic equilibria. For example the system $Mg^{2+} + AMP^{2-} \rightleftharpoons MgAMP$ can be used to buffer the free Mg^{2+} concentration. Clearly the buffer capacity of this system will be a maximum when $[AMP^{2-}] = [MgAMP]$.

Dissociation of polyprotic acids

So far we have considered acids which can lose only one proton (monoprotic acids). However, many biological species (e.g. nucleotides, nucleic acids, amino acids and proteins) are polyprotic acids. When alkali is added to such compounds a series of proton dissociations occurs. The simple case of the weak acid H_3PO_4 is illustrated in Fig. 7.2.

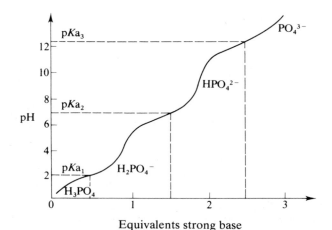

FIG. 7.2. Titration of 0.1 mol dm^{-3} phosphoric acid with a strong base showing the three distinct neutralization steps.

In this case the observed pK_a values† (2·0, 6·8, and 12·0) for the three acid species are sufficiently distinct to allow each separate step in the titration curve to be observed. From our previous discussion, we expect that each dissociation step has its own buffering region. For example, $H_2PO_4^-$ has a pK_a about 7 and so mixtures of $H_2PO_4^-$ and HPO_4^{2-} should be good buffers around neutral pH.

We could also consider the case where successive ionizations arise from different types of chemical groups, as for instance in amino acids. In glycine the pK_a of the carboxyl group is 2·34 and that of the amino group is 9·60. Thus at neutral pH, glycine exists as a 'zwitterion':

$$\overset{+}{N}H_3CH_2CO_2H \underset{2\cdot34}{\overset{pK_{a1}}{\rightleftharpoons}} \overset{+}{N}H_3CH_2CO_2^- \underset{9\cdot60}{\overset{pK_{a2}}{\rightleftharpoons}} NH_2CH_2CO_2^-$$

zwitterion

net charge
of each species   +1   0   −1

As glycine is titrated with strong alkali, there are two distinct proton dissociation steps. We would refer to glycine as an *amphoteric* electrolyte since it shows proton dissociations in both acidic and basic pH regions.

We would expect that at a certain pH the concentrations of positive and negative species are equal. This pH is known as the *isoelectric point*‡ and can be related to pK_{a1} and pK_{a2} as follows.

$$\overset{+}{N}H_3CH_2CO_2H \rightleftharpoons \overset{+}{N}H_3CH_2CO_2^- + H^+ \ldots (pK_{a1}),$$
$$\overset{+}{N}H_3CH_2CO_2^- \rightleftharpoons NH_2CH_2CO_2^- + H^+ \ldots (pK_{a2}).$$

Now

$$K_{a1} = \frac{[NH_3^+CH_2CO_2^-][H^+]}{[NH_3^+CH_2CO_2H]} \quad \text{and} \quad K_{a2} = \frac{[NH_2CH_2CO_2^-][H^+]}{[NH_3^+CH_2CO_2^-]},$$

$$\therefore \quad K_{a1} \cdot K_{a2} = \frac{[NH_3^+CH_2CO_2^-][H^+]}{[NH_3^+CH_2CO_2H]} \times \frac{[NH_2CH_2CO_2^-][H^+]}{[NH_3^+CH_2CO_2^-]},$$

---

† The actual pK_a values observed depend on the ionic strength of the solution (see p. 103). The true pK_a values (i.e. obtained by extrapolation to zero ionic strength) are 2·15, 7·2, and 12·4 respectively.

‡ Of course at this pH the solution will still conduct electricity but there will be no *net* migration in an electrophoresis experiment (where a voltage is applied across a solution). When the pH is less than the isoelectric point, the amino acid carries a net positive charge and will migrate towards the cathode. Above the isoelectric pH, the amino acid will migrate to the anode.

so that at the isoelectric point, where $[\overset{+}{N}H_3CH_2CO_2H] = [NH_2CH_2CO_2^-]$

$$K_{a1} . K_{a2} = [H^+]^2.$$

Taking logarithms we obtain

$$\log_{10} K_{a1} + \log_{10} K_{a2} = 2 \log_{10}[H^+]$$

or

$$pK_{a1} + pK_{a2} = 2 \, pH,$$

i.e. at the isoelectric point

$$\boxed{pH = \tfrac{1}{2}(pK_{a1} + pK_{a2})}.$$

The pH at the isoelectric point is often written as pI. For glycine $pI = \tfrac{1}{2}(2\cdot34 + 9\cdot60) = \underline{5\cdot97}$.

A slightly more complex example is that of glutamic acid, where the amino acid chain is acidic

$$HO_2\overset{\gamma}{C}-(CH_2)_2-CH \overset{\displaystyle \overset{+}{N}H_3 \quad (pK_{a3} = 9\cdot67)}{\underset{\displaystyle _{\alpha}CO_2H(pK_{a1} = 2\cdot2).}{\diagdown}}$$

$(pK_{a2} = 4\cdot25)$

Below pH 2, both carboxyl groups and the amino group are protonated, giving an overall charge of $+1$. When the pH is between $2\cdot5$ and 4, the α carboxyl group is ionized, while the γ-carboxyl group is hardly ionized, (overall charge $= 0$, i.e. the zwitterion). When the pH is between $4\cdot5$ and $9\cdot5$ both carboxyl groups are ionized (overall charge $= -1$). Finally above pH 10 the overall charge $= -2$.

Clearly the isoelectric point will be in the pH range where the zwitterion form is the major species. The exact pI is, as before, that pH at which the concentrations of singly positive and singly negative species are equal. For instance in this example $pI = \tfrac{1}{2}(pK_{a1} + pK_{a2}) = 3\cdot23$. The contribution of the doubly negative charged form at this pH is negligible, as illustrated in the example below. *In general, the pI is obtained by averaging the pK_a values of the groups which comprise the zwitterion form.*

Worked example

Show that the contribution of the doubly negative charged form of glutamic acid is negligible at pH $3\cdot23$, assuming that all the pK_a values are independent.

Solution

For the equilibrium below, $pK_a = 9·67$

$$^-O_2C-(CH_2)_2-CH\begin{array}{c}\overset{+}{N}H_3\\ \diagdown\\ CO_2^-\end{array} \rightleftharpoons {}^-O_2C-(CH_2)_2-CH\begin{array}{c}NH_2\\ \diagup\\ \diagdown\\ CO_2^-\end{array} +H^+$$

$$(glu\ \overset{+}{N}H_3) \qquad\qquad\qquad (glu\ NH_2)$$

Using the Henderson–Hasselbalch equation:

$$pH = pK_a - \log_{10}\frac{[glu\ \overset{+}{N}H_3]}{[glu\ NH_2]}$$

$$\therefore\quad 3·23 = 9·67 - \log_{10}\frac{[glu\ \overset{+}{N}H_3]}{[glu\ NH_2]}$$

$$\therefore\quad \frac{[glu\ \overset{+}{N}H_3]}{[glu\ NH_2]} = 2·76\times10^6,$$

i.e. only about 1 glutamic acid molecule in 10^6 exists as the doubly negative charged form at pH 3·23. Thus the contribution of this pK_a to the isoelectric point is negligible.

In proteins the situation is much more complex. Not only are there many different types of ionizing groups (glutamic acid, histidine, tyrosine, arginine etc.) but those of the same type may exist in different chemical environments and possess different pK_as. Although the interpretation of pH titration data is difficult, it is however possible to assign an isoelectric point to a protein (the pH at which the protein carries no net charge). For example thymohistone, which is rich in *basic* amino acids such as lysine and arginine, has a pI of 10·8 (cf. pI of lysine = 9·74, pI of arginine = 10·76). Serum albumen has a preponderance of *acidic* amino acids, such as glutamic acid and aspartic acid, and has a pI of about 4·8.

The effect of ionic strength on acid–base equilibria

In this chapter we have ignored the effect of non-ideality on the acid-base equilibria. However, in accurate work, we would consider the effect of ionic strength on the activity coefficients of the species involved in the equilibria (see Chapter 5, pp. 56–8). It is clear from the Debye–Hückel Law that deviations from ideal behaviour are more significant the more highly charged the ions involved (i.e. the higher the ionic strength). In the case of the dissociation of $0·1\ mol\ dm^{-3}$ HPO_4^{2-} to give PO_4^{3-} the observed pK_a (12) is markedly different from the true 'thermodynamic' pK_a of 12·4 for this dissociation (obtained by extrapolation to zero ionic strength).

The change in the activity coefficients of the ions with ionic strength is reflected in the change in the pH of a buffer solution on dilution. The change in pH on two fold dilution ($\Delta pH_{\frac{1}{2}}$) is +0·08 for $H_2PO_4^- - HPO_4^{2-}$ buffers and +0·05 for acetic acid–acetate buffers. Addition of a neutral salt (e.g. NaCl) to a buffer solution will also change the ionic strength and the effect of this on the pH of the solution is usually quite similar to that caused by increasing the concentration of the buffer solution.

PROBLEMS

1. The heat of neutralization of a strong acid by a strong base is $-56·8 \text{ kJ mol}^{-1}$. Calculate the ionic product (K_w) of water at 310 K, given that it is $0·61 \times 10^{-14}$ at 291 K. What is neutral pH at 310 K? ($R = 8·31 \text{ J K}^{-1} \text{ mol}^{-1}$.)

2. Calculate the pH of the following solutions, given $K_w = 10^{-14}$.
   (a) 0·05 mol dm^{-3} HCl
   (b) 0·1 mol dm^{-3} acetic acid ($K_a = 1·75 \times 10^{-5}$ (mol dm^{-3}))
   (c) 0·1 mol dm^{-3} aniline ($K_b = 3·82 \times 10^{-10}$ (mol dm^{-3}))
   (d) A mixture of 0·1 mol dm^{-3} acetic acid and 0·001 mol dm^{-3} HCl
   (e) 10^{-8} mol dm^{-3} HCl.

3. What is the activity coefficient of the hydrogen ion (γ_{H^+}) in a 0·01 mol dm^{-3} solution of HCl if the pH is 2·08?

4. What do you understand by the term 'isoelectric point'? What is the isoelectric point of:
   (a) Glyclglycine (p$K_{a_1} = 3·06$, p$K_{a_2} = 8·13$)
   (b) Aspartylglycine (p$K_{a1} = 2·1$, p$K_{a2} = 4·53$, p$K_{a3} = 9·07$)
   (c) Lysylglycine (p$K_{a1} = 2·1$, p$K_{a2} = 9·0$, p$K_{a3} = 10·6$)
   (d) Cysteine (p$K_{a_1} = 1·71$, pK_{a_2} (SH) = 8·33, pK_{a_3} (NH$_3^+$) = 10·78).

5. Derive the relationship between pH, pK and the concentration of acidic and basic forms of a compound, by taking logarithms of the appropriate equilibrium constant expression.
   (The relationship is often called the Henderson–Hasselbalch equation.)
   Use this equation to determine the volumes of 0·1 mol dm^{-3} NaH$_2$PO$_4$ and 0·1 mol dm^{-3} Na$_2$HPO$_4$ solutions which are required to produce 100 cm^3 of a buffer solution of pH 7·0.

   $$\text{For} \quad H_2PO_4^- \rightleftharpoons H^+ + HPO_4^{2-}, \quad pK_a = 7·2.$$

   What is the ionic strength of the two solutions and also of the final buffer?

6. What is meant by the term buffer solution? The ionization constant of acetic acid is $1·75 \times 10^{-5}$ (mol dm^{-3}) at 298 K. If a solution contains 0·16 moles of acetic acid per dm^3, how many moles of sodium acetate must be added to give a solution of pH 4·2. Discuss the effects on this solution which result from changes in (1) temperature and (2) the ionic strength.

7. By considering the neutralization of a solution of a weak acid by added hydroxide ion, it can be shown that the buffer value of the solution, i.e.

$$\frac{d \text{ (equivalents of base added)}}{d(\text{pH})}$$

is given by

$$\beta = 2·3 C\alpha(1-\alpha)$$

where β = buffer value
  C = concentration of weak acid
  α = degree of dissociation.
Use this equation to evaluate the pH values at which solutions of phosphate and tris act as most effective buffers.
  For $H_2PO_4^- \rightleftharpoons H^+ + HPO_4^{2-}$; $K_a = 6 \cdot 3 \times 10^{-8}$ (mol dm^{-3}) (298 K)
  for tris $H^+ \rightleftharpoons H^+ + $ tris; $K_a = 8 \cdot 3 \times 10^{-9}$ (mol dm^{-3}) (298 K).
  What is the ratio of buffer values of the following solutions to that of $0 \cdot 1$ mol dm^{-3} phosphate buffer at pH $7 \cdot 0$?
(a) $0 \cdot 1$ mol dm^{-3} phosphate at pH $8 \cdot 0$
(b) $0 \cdot 1$ mol dm^{-3} phosphate at pH $8 \cdot 5$
(c) $0 \cdot 01$ mol dm^{-3} phosphate at pH $7 \cdot 0$.
Comment on your answers.

8. ATP hydrolysis proceeds according to the following equation (at pH $8 \cdot 0$).

$$ATP^{4-} + H_2O \rightarrow ADP^{3-} + HPO_4^{2-} + H^+.$$

A 1 mmol dm^{-3} solution of ATP was hydrolyzed enzymatically in a $0 \cdot 1$ mol dm^{-3} tris buffer (pH $8 \cdot 0$). Calculate the new pH at the end of the reaction. What would have been the final pH if a $0 \cdot 01$ mol dm^{-3} tris buffer (pH $8 \cdot 0$) was employed and also if no buffer was used? (For tris, p$K_a = 8 \cdot 1$.)

9. The pH of a sample of arterial blood is $7 \cdot 50$. 20 cm^3 of this sample liberated $12 \cdot 2$ cm^3 of CO_2 at 298 K (1 atm.) on acidification (after correction for dissolved CO_2). If the pK_a for the reaction

$$CO_2 + H_2O \rightarrow H^+ + HCO_3^-$$

is $6 \cdot 1$, determine the concentrations of CO_2 and HCO_3^- in the blood. (1 mol CO_2 occupies $24 \cdot 45$ dm^3 at 298 K.)
  If the partial pressure of CO_2 (in atm) is related to the concentration of disolved CO_2 (in mol dm^{-3}) by the equation

$$[CO_2]_{dissolved} = 0 \cdot 031 \times P_{CO_2}$$

what is the pressure of CO_2 above the blood?

10. The following data were obtained for a titration of $0 \cdot 1$ mol dm^{-3} acetic acid with sodium hydroxide.

Equivalents NaOH added	0	0·10	0·20	0·40	0·60	0·80	0·90	0·95	1·00	1·05	1·10	1·30
pH	2·60	3·55	4·00	4·50	4·85	5·25	5·60	6·25	8·80	11·00	11·25	11·65

Explain which of the following indicators might be suitable for this titration:
Congo red (p$K_{In} = 4 \cdot 0$); Methyl red (p$K_{In} = 5 \cdot 2$);
Bromocresol purple (p$K_{In} = 6 \cdot 0$); Neutral red (p$K_{In} = 7 \cdot 4$);
Phenolphthalein (p$K_{In} = 9 \cdot 1$); Thymolphthalein (p$K_{In} = 9 \cdot 9$).

11. The mobility (m) of a peptide in electrophoresis (relative to aspartic acid) is believed to follow the relationship

$$m = KeM^{-\frac{2}{3}}$$

where K is a constant, e = charge (irrespective of sign), and M = molecular weight. At pH $6 \cdot 5$ the following results were obtained:

Peptide	Molecular weight	Mobility
Asp-Leu	246·3	0·65
Leu-Gly-Arg	344·5	0·53
Ileu-Ala-Ser-Lys-Phe	565	0·40
Asp-Gly-Asp	305·3	0·98
Gly-Arg-Lys	359·5	0·90
Asp-Gly-Leu-Asp	418·5	0·80

Plot $\log_{10}$(mobility) against $\log_{10}$(molecular weight) for the peptides of different charge classes.

Of the amino acids above, Asp carries a charge of -1 at pH 6·5, Arg and Lys a charge of $+1$ and the others (Leu, Gly, Ileu, Ala, Ser, Phe, Asparagine) carry no net charge.

Two peptides after acid hydrolysis were each found to contain only Asp and Leu in equimolar proportions. One peptide has a mobility of 0·75, and the other a mobility of 0·45. Use the above plot to deduce the probable formulae of the peptides.

(During acid hydrolysis the amide group of asparagine (Asp-NH$_2$) is converted to an acid group, yielding aspartic acid.)

12. If $\Delta H°$ for the uptake of a proton by tris-(hydroxymethyl) methylamine (tris) is $-46·0$ kJ mol^{-1}, what is the ratio of K_a at 298 K to that at 273 K? $\Delta H°$ for the uptake of a proton by the phosphate dianion is $-4·2$ kJ mol^{-1}. Calculate the change in K in this case and comment on the relative merits of these two buffers. ($R = 8·31$ J K^{-1} mol^{-1}.)

8. Electrochemical cells: oxidation–reduction processes

Oxidation–reduction processes

WE ARE devoting a separate chapter to a discussion of the thermodynamics of oxidation–reduction processes, not only to emphasize their importance in biochemistry, but also because many of these processes can be studied in electrochemical cells where true *reversible*† conditions can be achieved.

Oxidation–reduction processes can be most easily understood in terms of the transfer of electrons, e.g. the oxidation of Fe^{2+} to Fe^{3+}. One of the most important biochemical examples is the so-called 'mitochondrial electron transport chain', consisting of various cytochromes, flavoproteins etc. each of which can undergo oxidation and reduction. Electrons from reducing agents (such as NADH or succinate) are passed along the chain via these various cytochromes etc. to the oxidizing agent (in this case, O_2). In this way the free energy available from the oxidation of say, NADH by O_2 ($\Delta G^{o'} = -214 \cdot 4 \text{ kJ mol}^{-1}$) is utilized in a series of steps to produce three moles of ATP from ADP and phosphate. The ATP thus formed is used in a whole range of ways in the cell (e.g. mechanical work, biosynthesis, transport of metabolites etc.).

The thermodynamics of oxidation–reduction processes can be described by reference to a simple example.

Consider the reaction in which Zn is added to a solution of $CuSO_4$. Precipitation of Cu

$$Zn + Cu^{2+} \rightleftharpoons Zn^{2+} + Cu \downarrow,$$

occurs spontaneously, and under such conditions we would describe the reaction as being carried out irreversibly. For this reaction ΔG° is $-213 \cdot 2 \text{ kJ mol}^{-1}$ at 298 K (corresponding to an equilibrium constant of 10^{37}). The free energy available from this reaction could be utilized in the form of electrical work, by setting up an appropriate *electrochemical cell*. In order to do this we divide the overall reaction into two separate oxidation–reduction processes:

$$Zn - 2e^- \rightarrow Zn^{2+} \qquad \text{oxidation (i.e. loss of electrons),}$$

$$Cu^{2+} + 2e^- \rightarrow Cu \qquad \text{reduction (i.e. gain of electrons).}$$

(The overall reaction is obviously the sum of these two processes.)

The separate reactions are known as the *half-cell reactions*. Neither half-cell reaction could be studied in isolation, since the Zn would gain an

---

† The reader is referred to the discussion in Chapter 2.

enormous positive potential and the Cu a negative one. However, the difference between them can be studied if they can be suitably coupled to form an *electrochemical cell* (Fig. 8.1).

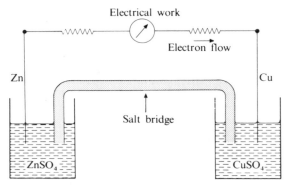

FIG. 8.1. An electrochemical cell to study the reaction $Zn + Cu^{2+} \rightleftharpoons Zn^{2+} + Cu$.

A suitable *electrochemical cell* to study this reaction is illustrated in Fig. 8.1. The electrodes (Zn and Cu) dip into solutions of their ions (the sulphate ions being necessary to maintain electrical neutrality). To avoid the spontaneous precipitation of Cu it is necessary to keep the solutions of $ZnSO_4$ and $CuSO_4$ separate, while maintaining electrical contact between them. This is achieved by a 'salt bridge', which contains concentrated KCl† in a gel and dips into each of the half cells.

If the Zn and Cu electrodes are connected, electrons flow from the Zn to the Cu as Zn^{2+} ions are formed and Cu^{2+} ions are reduced. As the reaction proceeds towards equilibrium the driving force (ΔG) falls and so the amount of electrical work obtained from the cell decreases.

Suppose however that we *apply* a suitable potential difference across the electrodes so as to oppose the effect of the electrochemical cell, as shown in Fig. 8.2.

As the applied potential difference is increased, the current flowing in the circuit will decrease. At a certain point (*the null point*), no current flows (i.e. the applied potential difference exactly equals that of the electrochemical

---

† KCl is chosen because K^+ and Cl^- have similar current carrying properties (mobilities), so that overall current can be carried in either direction with equal facility. The diffusion of Zn^{2+} or Cu^{2+} ions into the bridge is very slow and can effectively be ignored. However, it should be noted that a salt bridge (or any junction between liquids) contributes a small degree of irreversibility to the system. For most practical purposes, there is no need to consider this.

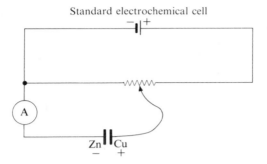

Zn/Cu electrochemical cell

FIG. 8.2. The Zn/Cu electrochemical cell with an opposing applied potential difference.

cell). Under these conditions the potential difference of the cell is known as its · *electromotive force* (*e.m.f.*). If the applied potential difference is increased beyond this value, the current will reverse its direction, (i.e. the cell reaction proceeds in the opposite direction $Cu + Zn^{2+} \rightarrow Cu^{2+} + Zn$). As the potential difference can be varied smoothly about the null point with no discontinuities in the current flow, the electrochemical cell is said to be operating *reversibly* at the null point. We have therefore used the applied potential difference to prevent the tendency of Zn to reduce Cu^{2+} ions in the cell reaction. A slight increase or decrease in the applied potential difference would drive the reaction in either direction, i.e. we have satisfied the criterion of *reversibility*. (Chapter 2.)

The thermodynamics of reversible cells

The First Law of thermodynamics states that

$$\Delta U = \Delta q + \Delta w,$$

where Δw is the work done *on* the system and consists of work done *by* the surroundings $(-P\Delta V)$ and also in this case the electrical work done $(-\Delta w_{elec})$.

So

$$\Delta U = \Delta q - P\Delta V - \Delta w_{elec}.$$

Now at constant pressure

$$\Delta H = \Delta U + P\Delta V \qquad \text{(see Chapter 1).}$$

So

$$\Delta H = \Delta q - \Delta w_{elec}.$$

Since the electrochemical cell process is made reversible, $\Delta q = T\Delta S$ (from the Second Law)

$$\therefore \quad \Delta H = T\Delta S - \Delta w_{\text{elec}}.$$

Using the equation $\Delta G = \Delta H - T\Delta S$ we obtain $\underline{\Delta G = -\Delta w_{\text{elec}}}$ (for electrochemical cell processes at constant pressure).

Thus ΔG measures the amount of 'useful' work obtainable from the electrochemical cell.

Suppose the reaction involves transfer of n electrons (e.g. $n = 2$ in the case of $Zn + Cu^{2+} \rightleftharpoons Zn^{2+} + Cu$). Then the electrical work done in transporting n electrons through a potential difference E is given by $\Delta w_{\text{elec}} = nFE$.

F is a conversion constant and is equal to $96 \cdot 5 \text{ kJ V}^{-1} \text{ mol}^{-1}$.

Thus we arrive at the equation

$$\boxed{\Delta G = -nFE}$$

for all reversible cell processes. Measurement of the e.m.f. E of an oxidation–reduction process gives ΔG directly, provided that the number of electrons n involved in this process is known. In the same way that ΔG° measures the free-energy change when reactants and products are in their standard states, E° refers to the *e.m.f. of the cell under standard state conditions*.

Worked example

For the reaction

$$Zn + Cu^{2+} \rightleftharpoons Zn^{2+} + Cu,$$

$$\Delta G^\circ = -213 \cdot 2 \text{ kJ mol}^{-1}.$$

What is the value of E° for this reaction?

Solution

$$\Delta G^\circ = -nFE^\circ,$$

$$\therefore \quad E^\circ = -\frac{\Delta G^\circ}{nF},$$

$$= +\frac{213 \cdot 2}{2 \times 96 \cdot 5} V \quad \text{since } n = 2,$$

$$= \underline{1 \cdot 1 \text{ V}}.$$

Note that since $\underline{\Delta G^\circ}$ is negative for reactions to be favourable, $\underline{E^\circ}$ must be positive.

Types of half-cells

As mentioned earlier all reactions in electrochemical cells can be broken down into two half cell reactions of the type

$$\text{Oxidized state} + ne^- \rightleftharpoons \text{Reduced state.}$$

(One of these reactions will be driven in reverse.)

We shall now illustrate the various types of half cell which are commonly used and indicate the nomenclature used to represent them.† Fig. 8.3 shows these type of half cells.

(1) *A metal electrode in a solution of its ions*

Many examples of this are known, e.g.

$$Zn^{2+} + 2e^- \rightleftharpoons Zn \qquad \ldots Zn^{2+}|Zn$$
$$Ag^+ + e^- \rightleftharpoons Ag \qquad \ldots Ag^+|Ag.$$

(2) *Gas in contact with a solution of its ions*

The best known example of this is the hydrogen electrode

$$H^+ + e^- \rightleftharpoons \tfrac{1}{2}H_2 \qquad \ldots H^+|\tfrac{1}{2}H_2 \cdot Pt.$$

Hydrogen gas is bubbled over a Pt black electrode, which dissociates the H_2 molecules. Thus a monatomic layer of hydrogen is formed in contact with the solution of H^+ ions.

Cl_2 and O_2 electrodes can also be used in this way.

(3) *Two different oxidation states of the same species*

Here an inert electrode (such as Pt) is in contact with the solution, and thus provides a means of allowing electrons to enter or leave the solution. For example

$$Fe^{3+} + e^- \rightleftharpoons Fe^{2+} \qquad \ldots Pt|Fe^{3+}, Fe^{2+}$$
$$Sn^{4+} + 2e^- \rightleftharpoons Sn^{2+} \qquad \ldots Pt|Sn^{4+}, Sn^{2+}$$
$$Fe(CN)_6^{3-} + e^- \rightleftharpoons Fe(CN)_6^{4-} \qquad \ldots Pt|Fe(CN)_6^{3-}, Fe(CN)_6^{4-}.$$

(4) *Metal in contact with its insoluble salt*

A metal is coated with a thin layer of one of its insoluble salts and this is in contact with a solution containing the anion of the insoluble salt. The two

---

† The convention used here is that a vertical line indicates the physical separation of two phases, e.g. Zn (solid) and Zn^{2+} solution ($Zn^{2+}|Zn$). A comma is used to indicate that the two species are in the *same* phase—e.g. Fe^{2+} and Fe^{3+} (Fe^{2+}, Fe^{3+}) in solution or Ag and AgCl both in the solid phase (Ag, AgCl). A double vertical line (‖) is used to denote a liquid–liquid junction (usually achieved by a salt bridge).

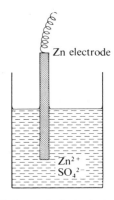

(1) A metal electrode in
a solution of its ions.

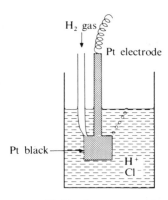

(2) Gas in contact with
a solution of its ions.

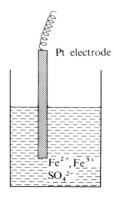

(3) Two different oxidation
states of the same species.

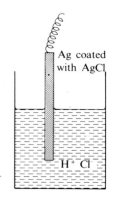

(4) Metal in contact with
its insoluble salt.

FIG. 8.3. The four common types of half-cell.

most common examples of this type are[†]

$$AgCl + e^- \; \rightleftharpoons \; Ag + Cl^- \qquad \ldots AgCl, Ag|Cl^-$$
$$\text{(solid)} \qquad\quad \text{(solid)}$$

and

$$\tfrac{1}{2}Hg_2Cl_2 + e^- \; \rightleftharpoons \; Hg + Cl^- \qquad \ldots \tfrac{1}{2}Hg_2Cl_2, Hg|Cl^-$$
$$\text{(solid)} \qquad\quad \text{(liquid)}$$

The latter is known as the *calomel electrode*, and is often used as a convenient reference electrode.

Half-cell electrode potentials

Clearly individual half-cell potentials cannot be measured directly as there is no way of measuring the potential difference between an electrode and the surrounding electrolyte. We can only measure the *difference* between the potentials of two half cells when they are linked to form an electrochemical cell.

A series of the relative values of electrode potentials could be obtained if each half cell were combined with a standard half cell. The hydrogen electrode has been chosen as this standard and assigned a standard e.m.f. ($E°$) of 0 volts, i.e. in the reaction

$$H^+ + e^- \; \rightleftharpoons \; \tfrac{1}{2}H_2(Pt)$$

when each component is in its standard state (H_2, 1 atm; H^+, a 1 mol dm^{-3} ideal solution) the e.m.f of this electrode = 0 volts. (This is the standard hydrogen electrode.)

Some representative *standard electrode potentials $E°$* for half cells (i.e. with each component in its standard state) are given in Table 8.1.

TABLE 8.1

Electrode	Reaction	$E°$ (V)
$Na^+|Na$	$Na^+ + \ e^- \to Na$	$-2\cdot71$
$Zn^{2+}|Zn$	$Zn^{2+} + 2e^- \to Zn$	$-0\cdot76$
$H^+|\tfrac{1}{2}H_2.Pt$	$H^+ + \ e^- \to \tfrac{1}{2}H_2(Pt)$	0
$Cu^{2+}|Cu$	$Cu^{2+} + 2e^- \to Cu$	$+0\cdot34$
$Pt|Fe^{3+}, Fe^{2+}$	$Fe^{3+} + \ e^- \to Fe^{2+}$	$+0\cdot77$
$Pt\cdot\tfrac{1}{2}Cl_2|Cl^-$	$(Pt)\tfrac{1}{2}Cl_2 + \ e^- \to Cl^-$	$+1\cdot36$

[†] The overall reaction in these half cells can be thought of as consisting of two simpler reactions, i.e.

$$AgCl \; \rightleftharpoons \; Ag^+ + Cl^-$$

and

$$Ag^+ + e^- \; \rightleftharpoons \; Ag.$$

The sign of $E°$ refers to the reaction as written in Table 8.1, i.e.

$$\boxed{\text{Oxidized form} + ne^- \;\rightarrow\; \text{Reduced form}}\,.$$

Thus a negative $E°$ value (i.e. $\Delta G°$ is positive) implies that the oxidized form is favoured. For instance H_2 will not reduce Zn^{2+} to Zn. Correspondingly a positive $E°$ (i.e. $\Delta G°$ is negative) means that the reduced state is favoured. Thus H_2 would reduce Cu^{2+} to Cu.

Electrochemical cells

Consider the linking of two half cells, for instance the Cu and Zn half cells, to form an electrochemical cell. We can represent this electrochemical cell as:†

$$Zn^{2+}|Zn\|Cu^{2+}|Cu.$$

It is *essential* to note that when the electrochemical cell is written in this way, it refers to the reaction as written below

$$Zn + Cu^{2+} \;\rightleftharpoons\; Zn^{2+} + Cu.$$

If the electrochemical cell were written

$$Cu^{2+}|Cu\|Zn^{2+}|Zn$$

this would refer to the reaction

$$Zn^{2+} + Cu \;\rightleftharpoons\; Zn + Cu^{2+}.$$

The *convention* is that the reaction corresponding to a given electrochemical cell,

$$(Ox)|(Red)\|(Ox)|(Red)$$

$$\text{Left} \qquad\qquad \text{Right}$$

is

$$\boxed{\text{Left (Red)} + \text{Right (Ox)} \;\rightleftharpoons\; \text{Left (Ox)} + \text{Right (Red)}}\,.$$

Worked examples

(1) What is the reaction corresponding to the electrochemical cell below?

$$Pt|Fe^{3+}, Fe^{2+}\|Sn^{4+}, Sn^{2+}|Pt.$$

What are the individual half cell reactions?

---

† Sometimes this type of cell is written with Zn on the extreme left hand side, i.e. $Zn|Zn^{2+}\|Cu^{2+}|Cu$. However the two representations refer to the same electrochemical cell. (See the convention referred to above.)

Solution

Using the above convention we see that the reaction is

$$2Fe^{2+} + Sn^{4+} \rightleftharpoons 2Fe^{3+} + Sn^{2+}.$$

Note that two electrons are involved in the reduction of Sn^{4+} to Sn^{2+}. Hence two equivalents of Fe^{2+} are required to maintain the electron balance.

The individual half-cell reactions are

$$Right \quad Sn^{4+} + 2e^- \rightleftharpoons Sn^{2+}$$

$$Left \quad 2Fe^{3+} + 2e^- \rightleftharpoons 2Fe^{2+}.$$

And the overall reaction is obtained by subtraction (Right–Left).

(2) Devise a cell to study the reduction of Co^{3+} to Co^{2+} by H_2 gas.

Solution

The reaction is

$$Co^{3+} + \tfrac{1}{2}H_2 \rightleftharpoons Co^{2+} + H^+.$$

Using the convention formula, the electrochemical cell would be

$$H^+|\tfrac{1}{2}H_2, Pt\|Co^{3+}, Co^{2+}|Pt.$$

Left   Right

The sign convention for electrochemical cells

The above examples illustrate that the overall reaction in an electrochemical cell is obtained by subtraction of the reaction at the *left-hand half-cell* from that at the *right-hand half-cell*.

Thus $E°$ for the electrochemical cell is given by

$$\boxed{E° \text{ (cell)} = E° \text{ (right)} - E° \text{ (left)}}.$$

Worked examples

(1) What is the $E°$ for the electrochemical cell?

$$H^+|\tfrac{1}{2}H_2, Pt\|Pt|Fe^{3+}, Fe^{2+}$$

given that

$$E° \text{ for } Pt|Fe^{3+}, Fe^{2+} \text{ is } +0.77 \text{ V}$$

and

$$E° \text{ for } H^+|\tfrac{1}{2}H_2, Pt \text{ is } 0 \text{ V}.$$

Solution

From the equation

$$E° \text{ (cell)} = E° \text{ (right)} - E° \text{ (left)}$$

we see that

$$\underline{E° \text{ (cell) is } +0·77 \text{ V.}}†$$

The cell reaction is

$$\tfrac{1}{2}H_2 + Fe^{3+} \rightleftharpoons H^+ + Fe^{2+}$$

and since $E°$ is positive (i.e. $\Delta G°$ negative) the above reaction would clearly proceed in the direction of Fe^{3+} reduction.

(2) What is the equilibrium constant K for the above reaction? (Assume $T = 298$ K.) ($R = 8·31$ J K^{-1} mol^{-1}.)

Solution

The electrochemical cell reaction involves the transfer of *one* electron (i.e. $n = 1$).
Using $\Delta G° = -nFE°$, $n = 1$, $F = 96·5$ kJ V^{-1} mol^{-1}, and $E° = +0·77$ V,

$$\therefore \quad \Delta G° = -1 \times 96·5 \times 0·77 \text{ kJ mol}^{-1}$$

i.e.

$$\Delta G° = -74·31 \text{ kJ mol}^{-1}.$$

Now

$$\Delta G° = -RT \ln K$$
$$\therefore \quad -74\,310 = -8·31 \times 298 \times \ln K$$

i.e.

$$\ln K = 30·00$$

and

$$\underline{K = 1·08 \times 10^{13} \text{ (mol dm}^{-3} \text{ atm).}}$$

This example illustrates how e.m.f. measurements can be used to determine $\Delta G°$ (and hence K) for a reaction. It is also possible to calculate $\Delta H°$ and $\Delta S°$ as illustrated below:

---

† This illustrates the general point that standard potentials are measured with respect to the hydrogen electrode when the latter is written to form the left hand half cell.

Calculation of thermodynamic parameters

Since $\Delta G° = \Delta H° - T\Delta S°$

$$\frac{d(\Delta G°)}{dT} = -\Delta S°$$

(assuming $\Delta H°$ is constant over the temperature range).

Now $\Delta G° = -nFE°$

$$\therefore \boxed{\frac{dE°}{dT} = \frac{\Delta S°}{nF}}.$$

So by determining $E°$ at two temperatures, $\Delta S°$ for the electrochemical cell reaction can be obtained. $\Delta H°$ is then readily calculated since

$$\Delta H° = \Delta G° + T\Delta S°.$$

So far we have only discussed the e.m.f.s of electrochemical cells under standard state conditions. We now derive a more general equation for the e.m.f. of an electrochemical cell under any given set of conditions.

The Nernst equation

Consider the equilibrium

$$a A + b B \rightleftharpoons c C + d D.$$

For this reaction, under standard state conditions $\Delta G°$ is given by

$$\Delta G° = c(G_C°) + d(G_D°) - a(G_A°) - b(G_B°).$$

Under any arbitrary non-standard conditions

$$\Delta G = c(G_C) + d(G_D) - a(G_A) - b(G_B).$$

So by difference

$$\Delta G - \Delta G° = c(G_C - G_C°) + d(G_D - G_D°) - a(G_A - G_A°) - b(G_B - G_B°).$$

Now we have to derive expressions for the terms $(G_C - G_C°)$, etc., for each component. The exact expression obviously depends on whether the components are gases, solutes or solids (Chapter 3).

For n moles of a gas $G - G° = nRT \ln (p/p°)$

For a solid $G = G°$

For a solute $G - G° = RT \ln (c/c°)$

where p and c represent pressure and concentration respectively. Note we are assuming ideal behaviour: if a solute, for example, is not behaving ideally we set its activity, $a = c\gamma$ where γ is the activity coefficient (Chapter 5).

All the terms could be combined in the form of a quotient. If all the components were solutes this would be

$$\Delta G - \Delta G^\circ = RT \ln \left(\frac{(c_C/c_C^\circ)^c (c_D/c_D^\circ)^d}{(c_A/c_A^\circ)^a (c_B/c_B^\circ)^b} \right)$$

Using $\Delta G = -nFE$, and remembering that $c_C^\circ = c_D^\circ$ etc. $= 1 \text{ mol dm}^{-3}$ we have

$$E = E^\circ - \frac{RT}{nF} \ln \frac{[C]^c [D]^d}{[A]^a [B]^b}$$

where [ ] represents concentration expressed in terms of mol dm^{-3}.

This is the *Nernst equation* written in terms of concentrations.

Let us consider some examples of the application of this equation.

Worked examples

(1) The standard electrode potential for the $Pt|Fe^{3+}, Fe^{2+}$ half cell is 0·77 V. What is the value of E for this half cell when $[Fe^{3+}] = 0·2 \text{ mol dm}^{-3}$ and $[Fe^{2+}] = 0·05 \text{ mol dm}^{-3}$? ($T = 298$ K.)

Solution

For the reaction below ($n = 1$)

$$Fe^{3+} + e^- \rightleftharpoons Fe^{2+}$$

the Nernst equation is

$$E = E^\circ - \frac{RT}{F} \ln \frac{[Fe^{2+}]}{[Fe^{3+}]}$$

assuming the solutions are ideal.

$$E = 0·77 - \frac{RT}{F} \ln \frac{0·05}{0·2}$$

$$= 0·77 - \frac{8·31 \, (298)}{96\,500} \ln (0·25)$$

$$= 0·77 + 0·036 \text{ V}$$

$$= \underline{0·806 \text{ V}}.$$

(2) What is the e.m.f. of the electrochemical cell

$$Pt, \tfrac{1}{2}H_2|H^+Cl^-|AgCl, Ag$$

if the concentration of HCl is 10^{-3} mol dm^{-3} and the partial pressure of H_2 is 1 atm? $E°$ for the cell is $0·2225$ V? ($T = 298$ K.)

Solution

The cell reaction is

$$\tfrac{1}{2}H_2 + AgCl \rightleftharpoons H^+ + Cl^- + Ag.$$
$$\text{(solid)} \qquad\qquad\qquad \text{(solid)}$$

For this reaction, $n = 1$.
The Nernst equation becomes

$$E = E° - \frac{RT}{F} \ln \frac{[H^+][Cl^-]}{(p_{H_2})^{\frac{1}{2}}}$$

since Ag and AgCl are both solids and hence in their standard state they do not appear in the expression above.

Now $[H^+] = [Cl^-] = 10^{-3}$ mol dm^{-3} and $p_{H_2} = 1$ atm.

$$E = 0·2225 - \frac{8·31(298)}{96\,500} \ln 10^{-6}$$

$$= 0·2225 + 0·3545 \text{ V}$$

$$E = \underline{0·577 \text{ V}}.$$

(3) What is the value of E for the hydrogen electrode if the partial pressure of $H_2 = 0·1$ atm? The concentration of H^+ is 1 mol dm^{-3}. ($T = 298$ K.)

Solution

For the reaction below, $n = 1$

$$H^+ + e^- \rightleftharpoons \tfrac{1}{2}H_2(g).$$

Now for the gas (H_2),

$$G - G° = nRT \ln p_{H_2}$$

and for the solute (H^+)

$$G - G° = RT \ln [H^+]$$

assuming the solution is ideal.

So the Nernst equation becomes

$$E = E° - \frac{RT}{F} \ln \frac{(p_{H_2})^{\frac{1}{2}}}{[H^+]}.$$

In this case

$$p_{H_2} = 0 \cdot 1, \qquad [H^+] = 1 \ mol \ dm^{-3}$$

$$E = 0 - \frac{8 \cdot 31(298)}{96\ 500} \ln (0 \cdot 1)^{\frac{1}{2}}$$

$$= 0 + 0 \cdot 0295 \ V$$

$$E = 0 \cdot 0295 \ V.$$

(4) What is the value of E for the half cell $Pt|NAD^+ + H^+$, NADH at pH 7, given that $E°$ is $-0 \cdot 11$ V? ($T = 298$ K). Assume that NAD^+ and NADH are both present at 1 mol dm^{-3} concentration.

Solution

The half-cell reaction for this most important biochemical oxidation–reduction system is:

$$\boxed{NAD^+ + H^+ + 2e^- \rightleftharpoons NADH} .$$

The Nernst equation is then

$$E = E° - \frac{RT}{2F} \ln \frac{[NADH]}{[NAD^+][H^+]}$$

Now $[H^+] = 10^{-7}$ mol dm^{-3} at pH 7.

$$\therefore \quad E = -0 \cdot 11 - \frac{RT}{2F} \ln \frac{1}{10^{-7}}$$

$$= -0 \cdot 11 - \frac{8 \cdot 31(298)}{2(96\ 500)} \ln 10^7$$

$$= -0 \cdot 11 - 0 \cdot 21 \ V$$

$$= -0 \cdot 32 \ V.$$

Thus at pH 7 the value of E for this half cell is $\underline{-0 \cdot 32 \ V}$.

Biochemical standard states

In Chapter 2 we introduced the concept of a 'biochemical standard state' where all the components of a reaction are present in their standard states, except H^+ which is present at 10^{-7} mol dm^{-3}, corresponding to pH 7. In the same way that the standard free energy change under these conditions is denoted by $\Delta G^{°\prime}$, we shall use the symbol $E^{°\prime}$ to refer to standard electrode potentials at pH 7. Thus in the example above $(Pt|NAD^+ + H^+$, NADH$)E^{°\prime} = -0 \cdot 32$ V.

(5) The $E^{\circ\prime}$ values for the reactions

$$\text{acetaldehyde} + 2H^+ + 2e^- \rightleftharpoons \text{ethanol} \qquad\qquad (i)$$

and

$$NAD^+ + H^+ + 2e^- \rightleftharpoons NADH \qquad\qquad (ii)$$

are $-0\cdot197$ V and $-0\cdot32$ V respectively.

What is the equilibrium constant for the coupled reaction?

$$\text{acetaldehyde} + NADH + H^+ \rightleftharpoons \text{ethanol} + NAD^+ \qquad (T = 298 \text{ K}).$$

Solution

$E^{\circ\prime}$ for the overall reaction is obtained by subtracting the $E^{\circ\prime}$ for reaction (ii) from that for reaction (i):

$$E^{\circ\prime} = -0\cdot197 - (-0\cdot32) \text{ V}$$

$$= +0\cdot123 \text{ V}.$$

Since

$$\Delta G = -nFE$$

$$\Delta G^{\circ\prime} = -2(96\cdot5)(0\cdot123) \text{ kJ mol}^{-1}$$

$$= -23\cdot7 \text{ kJ mol}^{-1}.$$

From

$$-\Delta G^{\circ\prime} = RT \ln K'$$

$$K' = 1\cdot43 \times 10^4 \text{ (mol dm}^{-3}).$$

Thus the equilibrium constant at pH 7 is $\underline{1\cdot43 \times 10^4 \text{ (mol dm}^{-3})}$.

The Nernst equation and chemical equilibrium

The above examples illustrate the use of the *Nernst equation* in relating the value of E° to the e.m.f. E under any arbitrary known set of conditions (e.g. concentrations). Values of E° for different electrochemical cells can be obtained and tabulated in the same way as values of ΔG°, etc.

It is a useful exercise to compare the Nernst equation with the equation linking ΔG° and equilibrium constant (i.e. $-\Delta G^\circ = RT \ln K$). It will be recalled that this latter equation was obtained as a special case of the more general relationship† (Chapter 3),

$$\Delta G - \Delta G^\circ = RT \ln \frac{[C]^c[D]^d}{[A]^a[B]^b}$$

---

† This is written in terms of concentrations, rather than activities.

since at equilibrium $\Delta G = 0$ and the values of [A], [B], etc., are their *equilibrium concentrations.*

However the above general equation can be used to calculate ΔG under any set of conditions (Chapter 3). In this respect, the general equation is entirely equivalent to the Nernst equation, which for solute components is:

$$E = E^\circ - \frac{RT}{nF} \ln \frac{[C]^c [D]^d}{[A]^a [B]^b}$$

(noting that $\Delta G = -nFE$).

In practice we usually set up an electrochemical cell with the components not in their standard states and measure E. The reaction is prevented from proceeding towards equilibrium by application of an opposing e.m.f. of magnitude E. If, however, we allowed the reaction to proceed (i.e. current to flow) the value of E would fall until, at equilibrium, E would equal zero (since at equilibrium $\Delta G = 0$).

The effect of non-ideality

So far we have assumed that all components in the electrochemical cells behave ideally. As we have seen in Chapter 5 this assumption can lead to errors especially in the case of electrolyte solutions. As an example consider the cell

$$\text{Pt}, \tfrac{1}{2}\text{H}_2(1 \text{ atm})|\text{H}^+\text{Cl}^-(\text{conc.} = c)|\tfrac{1}{2}\text{Hg}_2\text{Cl}_2, \text{Hg}.$$

The cell reaction is

$$\tfrac{1}{2}\text{H}_2 + \tfrac{1}{2}\text{Hg}_2\text{Cl}_2 \rightleftharpoons \text{Hg} + \text{H}^+ + \text{Cl}^-$$

and the Nernst equation,

$$E = E^\circ - \frac{RT}{F} \ln [\text{H}^+][\text{Cl}^-] \qquad (n = 1)$$

can be used to calculate the apparent E° as a function of the HCl concentration(c). The following results are obtained.

c(M)	E(V)	E°(V)
10^{-1}	0·4046	0·2866
10^{-2}	0·5099	0·2739
10^{-3}	0·6239	0·2699
10^{-4}	0·7406	0·2686

E° is not constant because of the non-ideality of the solution. It is possible to account for this departure from ideality by substituting activities ($a = c\gamma$) for concentrations in the Nernst equation and calculating γ from the Debye–Hückel theory for dilute electrolyte solutions (Chapter 5). When this is done E° has a constant value (0·2680 V), and this would be the true E° obtainable by extrapolation of the above E° values to zero ionic strength (i.e. infinite dilution).

Coupled oxidation–reduction processes

From our discussions of the thermodynamics of oxidation–reduction processes it is clear that e.m.f. values can be thought of as entirely analogous to free energy changes. Now we can use ΔG° values for individual reactions to predict the equilibrium position in a 'coupled' reaction and similarly in a 'coupled' oxidation–reduction process,

$$(Ox)_1 + (Red)_2 \rightleftharpoons (Ox)_2 + (Red)_1$$

we can use E° values to predict the position of equilibrium ($\Delta G^\circ = -nFE^\circ$) provided of course n is known. One of the most important examples of such processes is the mitochondrial 'electron transport chain', where a series of 'half-cell reactions' are arranged in order of increasing $E^{\circ\prime}$ (i.e. stronger oxidizing power). Electrons are passed along the chain, and the total energy available from oxidation of NADH is utilized (by a process not yet fully understood) in the production of ATP from ADP and phosphate. A simplified energy diagram, showing some of the more important components of the chain is given in Fig. 8.4. Two electrons are assumed to be passed along the chain so that from $\Delta G^{\circ\prime} = -nFE^{\circ\prime}$, we see that for each 0.1 V change in $E^{\circ\prime}$ the $\Delta G^{\circ\prime}$ change is 19.3 kJ mol^{-1}.

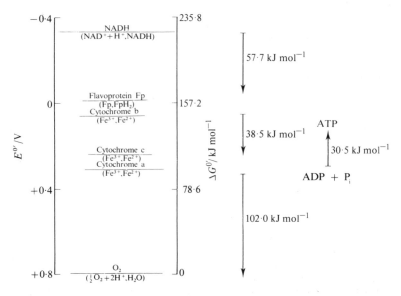

F IG. 8.4. Some of the more important components of the mitochondrial electron-transport chain. The $\Delta G^{\circ\prime}$ values are relative to the oxygen, water couple as zero.

There are three places indicated in the chain where the difference in $\Delta G^{o\prime}$ is sufficient to drive the phosphorylation of ADP to produce ATP ($\Delta G^{o\prime}$ for this process is $30\cdot5$ kJ mol^{-1}), and these three places have been tentatively described as the 'phosphorylation' sites. It should be remembered, however, that this assumes that the various components are all in their standard states, and clearly the *in vivo* situation (where the components are not present in their standard states, and the $\Delta G'$ values differ considerably from the $\Delta G^{o\prime}$ values) will be considerably more complicated than this.

Oxidative phosphorylation in the mitochondria

The mechanism of coupling of electron transport to ATP synthesis has been difficult to unravel. There is now a general, but not universal, consensus that protons play a key role in the coupling mechanism and that a *chemiosmotic hypothesis* gives a very useful description of the mechanism. In this hypothesis, the electron flow from one electron carrier to another at a higher redox potential is coupled to the movement of protons from the inside (matrix) to the outside of the mitochondrion. This hypothesis is the only one which is at present amenable to a quantitative treatment and we therefore develop the necessary thermodynamic treatment.

We first have to introduce a new thermodynamic function, the *electrochemical potential* of an ion. We have already briefly mentioned the concept of *chemical potential* (Chapter 5) of a solute, μ, which is essentially equivalent to its molar Gibbs free energy. For an ideal solution the *chemical* potential of an ion, μ_i, can be defined as

$$\mu_i = \mu_i^o + RT \ln X_i$$

where μ_i^o is the chemical potential of the pure component in its standard state (cf. eqn (5.5)). For an ion, however, the chemical potential, μ_i, alone is not the correct term to describe its thermodynamic behaviour since we also have to take account of the interaction of the ion with the electrostatic field, ψ, in which it is placed. This field is provided by the neighbouring charged species. This additional term equals $zF\psi$ where z is the charge on the ion and F is the conversion factor ($96\cdot5$ kJ V^{-1} mol^{-1}).

The *electrochemical potential*, $\tilde{\mu}_i$, is therefore defined as

$$\tilde{\mu}_i = zF\psi + \mu_i$$
$$= zF\psi + \mu_i^o + RT \ln X_i. \tag{8.1}$$

In the chemiosmotic hypothesis, the movement of protons from the matrix to the outside of the mitochondria will cause† the inside (matrix) of the mitochondria to become negatively charged and alkaline and the outside to become positively charged and acidic. There is now energy stored as a gradient across the membrane (which is highly impermeable to protons) and this energy *is* the difference in electrochemical potential for the hydrogen ion. Therefore, from eqn (8.1)

$$\tilde{\mu}_{i\,in} - \tilde{\mu}_{i\,out} = zF\psi_{in} - zF\psi_{out} + RT \ln \left(\frac{[\mathrm{H}^+]_{in}}{[\mathrm{H}^+]_{out}} \right)$$

or

$$\Delta\tilde{\mu}_i = zF\Delta\psi - 2\cdot3\,RT\,\Delta\mathrm{pH}$$

† Neglecting any buffering effects and provided that no other ions are moving.

where ΔpH is the pH difference between the inside and outside and $\Delta\psi$ is the membrane potential across the membrane (inside minus outside). It is convenient to use units of mV for the electrochemical potential and the value of $\Delta\bar{\mu}_i$ in mV is termed the *protonmotive force* Δp. To convert the terms in the above equation to electrical units (i.e. volts) we divide through by the conversion factor F. Putting $\Delta\bar{\mu}_i/F = \Delta p$ and noting that $z = 1$ for the proton, we obtain

$$\Delta p = \Delta\psi - \frac{2 \cdot 3RT}{F}\Delta pH \quad .\dagger \tag{8.2}$$

$\Delta\psi$ is the membrane potential (in mV).

We now consider two very important questions.

(1) How large does the protonmotive force have to be to drive oxidative phosphorylation?
(2) How is the protonmotive force measured?

We shall consider each in turn.

(1) *The magnitude of the protonmotive force*

Nothing has been said so far on the question of *how many* protons are translocated as electrons pass down the respiratory chain. This is, at present, a controversial topic. However, it should be noted that the driving force required to move the proton(s) (which is provided by the redox potential difference between two adjacent carriers) is directly proportional to the number of protons being moved.

How does this protonmotive force drive ATP synthesis? The proposal is that ATP synthesis occurs when a defined number of protons pass down their electrochemical gradient via mitochondrial ATPase‡ (see Fig. 8.5). It follows that the protonmotive force must be large enough to favour ATP synthesis from ADP and P_i. The required magnitude of Δp could be assessed in the following experiment. A suspension of mitochondria is supplied with oxidisable substrate, ADP and P_i, and allowed to respire until the ATP concentration reaches a steady value. The final concentrations of ADP, P_i, and ATP are then measured. This enables the $\Delta G'$ for the reaction $ADP + P_i = ATP + H_2O$ to be calculated provided that $\Delta G^{o\prime}$ is also known since

$$\Delta G' = \Delta G^{o\prime} + RT \ln \frac{[ATP]}{[ADP][P_i]} \tag{8.3}$$

$\Delta G'$ may then be related to the protonmotive force by the equation we have

---

† To obtain the quantities in mV, we should note that the value of F is $96 \cdot 5 \text{ J mV}^{-1} \text{ mol}^{-1}$.

‡ The membrane is highly impermeable to protons and it is only via this ATPase the protons can move back across the membrane into the matrix.

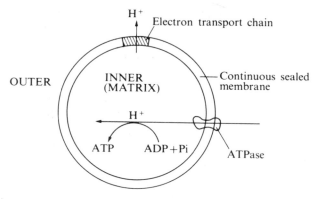

FIG. 8.5. Representation of oxidative phosphorylation in the mitochondrion according to the chemiosmotic hypothesis.

already met, $\Delta G' = nFE'$. Thus

$$\Delta G' = -n \, . \, F \, . \, \Delta p \qquad\qquad (8.4)$$

where n is the number of protons that pass through the ATPase for each ATP molecule synthesized. In the *chemiosmotic hypothesis the value of n* = 2 and the required value of Δp can be calculated.

(2) *Measurement of the protonmotive force*

We now require a method of estimating the magnitude of the proton-motive force so that the thermodynamic feasibility of the chemiosmotic hypothesis can be tested. Eqn (8.2) shows that we need to know ΔpH and the membrane potential $\Delta\psi$ to obtain Δp. We shall consider each in turn.

(i) *Measurement of ΔpH*

The most widely used method for estimating ΔpH involves the use of a metabolically inert weak base or acid. Biological membranes are generally not permeable to charged molecules and thus the underlying assumption made is that only the uncharged form, of the acid or base, can cross the membrane.

Worked example—determination of ΔpH

A $1\,cm^3$ suspension of rat liver mitochondria in $0.25\,mol\,dm^{-3}$ sucrose pH 6·0 is incubated with succinate and [^{3}H]-acetate for 2 minutes before the mitochondria were very rapidly separated from the incubation medium.

Analysis showed that one sixth of the total acetate had been taken up by the mitochondria. If the concentration of mitochondria was 5 g protein dm^{-3}, what was the magnitude of the pH gradient? Assume that the internal volume of mitochondria is $0\cdot4$ cm^3 (g protein)$^{-1}$.

Solution

Assume that only CH$_3$COOH (denoted HA) but not CH$_3$COO$^-$ (denoted A$^-$) can permeate the membrane. Then at equilibrium

$$[HA]_{in} = [HA]_{out}$$

but

$$K_a = \frac{[H^+]_{in}[A^-]_{in}}{[HA]_{in}} = \frac{[H^+]_{out}[A^-]_{out}}{[HA]_{out}}$$

so that

$$[H^+]_{in}/[H^+]_{out} = [A^-]_{out}/[A^-]_{in}.$$

However the measurements only tell us the total acetate $[A^-_{in} + AH_{in}]$ inside the mitochondria. Fortunately we only need to know the total concentration as the following will show:

$$\frac{[Acetate]_{total_{in}}}{[Acetate]_{total_{out}}} = \frac{[AH]_{in} + [A^-]_{in}}{[AH]_{out} + [A^-]_{out}}$$

$$= \frac{[AH]_{in} + (K_a([AH]_{in}/[H^+]_{in}))}{[AH]_{out} + (K_a([AH]_{out}/[H^+]_{out}))}.$$

But we have already noted that $[AH]_{in} = [AH]_{out}$, so

$$\frac{[Acetate]_{total_{in}}}{[Acetate]_{total_{out}}} = \frac{1 + K_a/[H^+]_{in}}{1 + K_a/[H^+]_{out}}$$

Dividing by K_a on the right hand side we obtain

$$\frac{[Acetate]_{total_{in}}}{[Acetate]_{total_{out}}} = \frac{1/K_a + 1/[H^+]_{in}}{1/K_a + 1/[H^+]_{out}}.$$

If the pK_a of the acid is $\ll$ pH$_{in}$ or pH$_{out}$ (i.e. if K_a for the acid is greater than $[H^+]_{out}$ or $[H^+]_{in}$) then:

$$\frac{[Acetate]_{total_{in}}}{[Acetate]_{total_{out}}} = \frac{[H^+]_{out}}{[H^+]_{in}}.$$

Since 1/6th of the acetate was taken up into the mitochondria 5/6th remain on the outside in the suspending medium. The relative volumes of the mitochondria and the suspending medium are known so that the acetate concentration ratio in the above equation can be obtained as follows. The internal volume is 0.002 cm^3 (0.0004×5) while the total volume is 1 cm^3. The relative concentrations of acetate are then obtained by dividing the relative amounts of acetate in each compartment by the corresponding volume. Thus

$$\frac{[\text{Acetate}]_{\text{total in}}}{[\text{Acetate}]_{\text{total out}}} = \frac{\frac{1}{6} \div 0.002}{\frac{5}{6} \div 1} = \frac{100}{1}$$

$$\therefore \quad [H^+]_{\text{out}}/[H^+]_{\text{in}} = 100.$$

Since the external pH is 6.0, the internal pH is estimated to be 8.0, i.e. the inside is more alkaline with $\Delta pH = 2.0$.

The distribution of a weak acid or base is a method which has found wide application in biology for the estimation of the pH in cellular spaces. The reader should note that the method involves a number of assumptions which cannot always be justified by experiment. Amongst these are the assumptions of equal activity coefficients on each side of the membrane, no binding or metabolism of the weak acid or base, and that complete equilibration via the uncharged form has occurred. It should also be noted that the accumulation of a species is always easier to measure than its exclusion and so if the organelle under investigation is expected to have a high pH relative to the medium then a weak acid probe should be used. Conversely, when a more acidic internal pH is suspected the distribution of a weak base should be measured.

(ii) Measurement of membrane potential

The membrane potential, $\Delta\psi$ is estimated on the basis of an ion distribution across a membrane. To understand the method we recall the definition of the electrochemical potential of an ion, eqn (8.1). The expressions for the difference $\Delta\tilde{\mu}_i$ in electrochemical potentials that can occur across a membrane boundary is

$$\Delta\tilde{\mu}_i = zF\Delta\psi + RT \ln \frac{[X_i]_{\text{in}}}{[X_i]_{\text{out}}}. \tag{8.5}$$

Let us consider how this equation can be applied to the behaviour of a positively charged ion to which the mitochondrial membrane is permeable. (The membrane becomes permeable to certain hydrophobic organic cations and to certain inorganic ions such as K^+ or Rb^+ when an ionophore, which is an 'ion carrier' molecule, is present.) From the earlier description of the chemiosmotic hypothesis we expect respiratory chain-linked redox reactions to make the outside of the mitochondria positive, relative to the inside, following the translocation of H^+ ions. The positively charged ion (e.g. K^+ or Rb^+) will enter the mitochondria until it reaches a distribution across the mitochondrial membrane which is in electrochemical equilibrium, i.e. $\Delta\tilde{\mu}_i = 0$. When this condition is satisfied, then from eqn (8.5)

$$\Delta\psi = -\frac{RT}{zF}\ln\left(\frac{[X]_{in}}{[X]_{out}}\right). \tag{8.6}$$

Thus if $[X]_{in}$ and $[X]_{out}$ can be measured, $\Delta\psi$ can be estimated.

Knowing $\Delta\psi$ and ΔpH, Δp can then be determined from eqn (8.2).

We have attempted to provide the necessary thermodynamics for a basic understanding of the chemiosmotic hypothesis as applied to mitochondria. The same hypothesis is in principle applicable to energy linked transport processes in chloroplasts and bacteria. A cautionary note is apposite at this point. A large amount of qualitative experimental evidence has been used to support the chemiosmotic hypothesis—but a mechanism can only be established by thermodynamic (and kinetic—see Chapter 9) measurements. We have outlined some of the thermodynamic measurements which involve a number of assumptions which may not be satisfied. Research workers are still actively concerned with the mechanism of oxidative phosphorylation and it may yet transpire that the protonmotive force has been either over- or underestimated by these measurements. At present however, the chemiosmotic hypothesis provides the only framework for a quantitative thermodynamic analysis of the coupling of electron transport to ATP synthesis.

PROBLEMS

1. What are the reactions in the following half cells?
   (a) $Zn^{2+}|Zn$
   (b) $H^+|\frac{1}{2}H_2.Pt$
   (c) $Pt|Co^{3+}, Co^{2+}$
   (d) $AgBr, Ag|Br^-$
   (e) $\frac{1}{2}Hg_2Cl_2, Hg|Cl^-$ (the calomel electrode).
   (f) $Pt|Fumarate^{2-} + 2H^+$, succinate $^{2-}$
   (g) $Pt|Cytochrome\ c\ (Fe^{3+})$, Cytochrome c (Fe^{2+})
   (h) $Pt|CO_2 + H^+$, Formate$^-$.
   (i) $Pt|NAD^+ + H^+$, NADH.

2. Write down the cell reaction for each of the following cells:
   (a) $Cu^{2+}|Cu\|Zn^{2+}|Zn$
   (b) $H^+|\frac{1}{2}H_2.Pt\|Ag^+|Ag$
   (c) $Pt.\frac{1}{2}H_2|HCl|AgCl, Ag$
   (d) $H^+|Pt.\frac{1}{2}H_2\|Fe^{3+}, Fe^{2+}|Pt$
   (e) $Pt|NAD^+ + H^+$, NADH$\|$oxaloacetate$^{2-} + 2H^+$, malate$^{2-}|Pt$.
   Comment on the involvement of H^+ ions in (e). Would you write this cell differently?
   Calculate the standard e.m.f.s of the above cells $(a)-(d)$ given the following standard electrode potentials (in volts) all at pH = 0: $Cu^{2+}|Cu$, 0·34; $Zn^{2+}|Zn$, $-0·76$; $H^+|\frac{1}{2}H_2.Pt = 0$; $Ag^+|Ag = 0·80$; $AgCl, Ag|Cl^- = 0·22$; $Pt|Fe^{3+}, Fe^{2+}$, $+0·77$.

For cell (e) $E°$ values at pH 7 (known as $E°'$—by analogy with $\Delta G°'$) are: Pt|NAD$^+$+H$^+$, NADH, -0.32 V; Pt|Oxaloacetate^{2-}+2H$^+$, malate^{2-}, -0.17 V.

Calculate $E°'$ for the cell.

3. Write down the cells equivalent to the following reactions:
   (a) Sn^{2+}+Pb $\rightleftharpoons$ Pb^{2+}+Sn
   (b) Lactate$^-$+NAD$^+$ $\rightleftharpoons$ pyruvate$^-$+NADH+H$^+$.

4. Discuss the meaning of the term *reversible* as used in the second law of thermodynamics.

   In the reaction

$$Fe+Cu^{2+} \rightleftharpoons Fe^{2+}+Cu,$$
$$\Delta H_{298}^\circ = -148.8 \text{ kJ mol}^{-1}$$

and

$$\Delta S_{298}^\circ = 8.8 \text{ J K}^{-1} \text{ mol}^{-1}.$$

Under standard state conditions:
   (a) In which direction would the reaction proceed?
   (b) In a cell an e.m.f. is applied to prevent this reaction occurring—calculate this e.m.f. (298 K).
   (c) Discuss *briefly* the situation when the applied e.m.f. is different from this value.

5. The standard e.m.f. of the cell

$$Zn^{2+}|Zn\|Fe^{3+}, Fe^{2+}|Pt$$

is 1.53 V at 298 K and 1.55 V at 323 K.
   What is the cell reaction?
   Calculate $\Delta G°$, $\Delta H°$, and $\Delta S°$ and state any assumptions you make.

6. Derive the Nernst equation for a half cell. Using this calculate the ratio of oxidized to reduced forms of cytochrome c_1 at the following potentials: 0.3, 0.25, 0.2, 0.15, and 0.1 volts (T = 298 K)

$$Pt|Cyt\text{-}c_1 (Fe^{3+}), Cyt\text{-}c_1 (Fe^{2+}) \qquad (E°' = 0.21 \text{ V})$$

7. (a) For the NAD$^+$+H$^+$, NADH couple, the value of $E°'$ (i.e. at pH = 7) is -0.32 V. What is the value of the e.m.f. at pH = 0 (i.e. $E°$)?
      If both NAD$^+$ and NADH are in their standard state (i.e. 1 mol dm^{-3}) but the pH = 6, what is the new e.m.f.?
   (b) For the oxaloacetate^{2-}+2H$^+$, malate^{2-} couple, $E°' = -0.175$ V (i.e. pH = 7). What is the e.m.f. at pH = 6, if both anions are in their standard state?
   (c) From the data in (a) and (b) calculate the value of the equilibrium constant for the oxidation of malate^{2-} by NAD$^+$ at pH = 6 and pH = 7. Comment on your answer. Assume the temperature is 298 K.

8. The pH of a solution could be measured by using a standard hydrogen electrode as in the following cell:

$$Pt.\tfrac{1}{2}H_2(1 \text{ atm})|HCl(x \text{ mol dm}^{-3})|\tfrac{1}{2}Hg_2Cl_2, Hg.$$

This cell has an e.m.f. of 0.48 V at 298 K. What is the pH of the solution? ($E°$ for the calomel electrode is 0.24 V.)

9. In the photosynthetic chain the following oxidation–reduction couples are involved:

$$\text{Cyt-b (Fe}^{3+}), \text{ Cyt-b (Fe}^{2+}), \qquad E^{o\prime} = 0\cdot06 \text{ V}$$

$$\text{Cyt-f (Fe}^{3+}), \text{ Cyt-f (Fe}^{2+}) \qquad E^{o\prime} = 0\cdot36 \text{ V}$$

What is the $\Delta G^{o\prime}$ of the overall reaction in which these are coupled? Is this energy sufficient to synthesize ATP from ADP and P_i, if $\Delta G^{o\prime}$ for this latter reaction is $+30\cdot5$ kJ mol^{-1}?

Would the same conclusion be reached if we considered the flow of two electrons along the redox chain, i.e.

$$2 \text{ Cyt-b (Fe}^{3+}) + 2\text{e}^- \rightleftharpoons 2 \text{ Cyt-b (Fe}^{2+})$$

$$2 \text{ Cyt-f (Fe}^{3+}) + 2\text{e}^- \rightleftharpoons 2 \text{ Cyt-f (Fe}^{2+})?$$

10. A 1 cm^3 suspension of mitochondria (5 g protein dm^{-3}) was incubated with $[^3\text{H}]$-acetate, $^{86}\text{Rb}^+$, valinomycin (ionophore which makes the membrane permeable to Rb^+), ADP, P_i and succinate, for a sufficient period to permit the ATP level to reach a constant value. At the end of this incubation the mitochondria were rapidly filtered from $0\cdot5$ cm^3 of the suspension and counted for radioactivity from $[^3\text{H}]$-acetate and $^{86}\text{Rb}^+$. It was found that the amount of $[^3\text{H}]$-acetate in the mitochondria was $1/11$th of the total acetate added, while $1/6$ of the total $^{86}\text{Rb}^+$ was also found in the mitochondria. The remaining $0\cdot5$ cm^3 of the mitochondria suspension was analysed for ATP, ADP, and P_i. These concentrations were 1 mmol dm^{-3}, $0\cdot02$ mmol dm^{-3}, and 1 mmol dm^{-3} respectively. Are these data compatible with the view that 2 protons must pass across the mitochondrial membrane for each ATP molecule synthesized? Assume that the internal volume of mitochondria is $0\cdot4$ cm^3 g^{-1} protein. ($T = 300$ K, $F = 96\cdot5$ kJ V^{-1} mol^{-1}. $R = 8\cdot31$ J K^{-1} mol^{-1}. $\Delta G^{o\prime}$ for the reaction $\text{ADP} + P_i = \text{ATP}$ is $30\cdot5$ kJ mol^{-1}.)

9. Chemical kinetics

The pathway of a reaction

THE study of thermodynamics enables us to predict which processes or reactions might occur spontaneously. However, it cannot tell us about the rate at which such processes occur, and it is this aspect which is encompassed in the subject of *kinetics*. For example, the hydrolysis of ATP to give ADP and phosphate is known to be highly favourable thermodynamically ($\Delta G^{o\prime} = -30 \cdot 5$ kJ mol^{-1} at pH 7·0 and 298 K), yet a solution of ATP at pH 7·0 is fairly stable. We can explain this *kinetic stability* of ATP by reference to Fig. 9.1 which illustrates a typical 'energy profile' for a reaction:

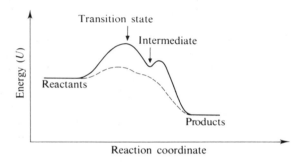

FIG. 9.1. The pathway of a reaction. The solid line represents energy changes during the course of a hypothetical reaction. The broken line refers to a possible path when a catalyst is present.

The solid line indicates the energy content as the reaction proceeds. The term '*reaction coordinate*' is plotted as the *x*-axis, and can be thought of as a 'measure of the extent of reaction', although its precise meaning is difficult to define. For instance, in the case of ATP hydrolysis, it could be taken to refer to the distance between the P and O atoms forming the bond which is being broken, indicated by the zig-zag line.

Breaking such a bond would require the input of a considerable amount of energy and it is essentially this 'energy barrier' which is responsible for the stability of ATP. The term *transition state* refers to the maximum in the energy curve, and if there is a minimum this is termed an *intermediate*. The difference between the energy of the reactants and that of the highest energy transition state is known as the *activation energy* of the reaction. A catalyst would lower this activation energy and hence speed up the reaction, though it would not cause any change in the energy levels of the reactants and products† (this is shown by the dotted line in Fig. 9.1). In the case of ATP hydrolysis, the catalyst might be an enzyme (ATPase).

Fig. 9.1 also illustrates the basic difficulty in the study of a reaction *mechanism* (i.e. the pathway by which a reaction proceeds). We are interested in the properties of the transition state and yet, by its very nature, this is an unstable state. If we can isolate an intermediate (and this depends on how marked the dip in the energy curve is) this may give us some clue as to the reaction pathway, but the study of transition states clearly requires an indirect approach. By contrast thermodynamics is a more exact subject because it deals only with the properties of the reactants and products independent of the pathway by which they are interconverted.

Kinetics is a subject in which the rate of a reaction is studied as certain parameters are varied; such as the concentration of reactants, temperature, pressure, pH, etc. The data accumulated is then analysed so as to give certain *rate laws*, and hence to yield some information about the *mechanism* of the reaction. The postulated mechanism (usually consisting of the elementary steps involved in the reaction, with some idea of their rates and energy changes) can then be used as a basis for predicting other features of the reaction (such as the dependence of its rate on ionic strength), which are open to experimental test.

We shall now examine certain aspects of the kinetics of reactions. The first point is the *reaction order*.

The order and molecularity of a reaction

The order is defined as the power to which the concentration of a reactant is raised in the *rate law*. Thus in the reaction below where x moles of A combine with y moles of B

$$x A + y B \rightarrow \text{products}$$

the expression for the rate of formation of product (P), $d[P]/dt$, might be

$$\frac{d[P]}{dt} = k[A]^a[B]^b,$$

where [ ] represents concentration, and k is a constant known as the *rate constant*.

---

† This means that the catalyst does not change the position of equilibrium of the reaction, only the rate of the reaction.

The reaction would then be ath order in A, bth order in B, with an overall order $(a + b)$.†

The *order* of a reaction is an experimental quantity and can have integral values $(0, 1, 2, \ldots)$ or non-integral values. For instance the decomposition of gaseous acetaldehyde to give CO and CH_4 is of approximately three halves order in acetaldehyde, and many enzyme-catalysed reactions are of a nonintegral order in substrate in certain ranges of substrate concentration. (See Chapter 10 on *Enzyme kinetics*.)

Order should not be confused with *molecularity*. The molecularity of a reaction is the minimum number of species involved in the slowest (or rate-determining) step of the reaction. Reference to the energy profile diagram shows that this is the same as saying that the molecularity is the minimum number of species involved in the transition state. We see that molecularity depends on the proposed mechanism of the reaction and thus cannot be deduced from a simple experiment (as can *order*). It is also clear that the molecularity of a reaction must be integral, whereas the order is often non-integral.

Various types of rate law can now be considered.

Types of rate processes

(1) *Zeroth-order processes*

In this case the rate of the reaction does not depend on the concentration of the reactant A.

$$-\frac{d[A]}{dt} = k.$$

On integration $[A]_0 - [A]_t = kt$, where $[A]_0$ is the initial value of $[A]$, and $[A]_t$ is the value at time t. Thus the concentration of A falls in a linear fashion with time (Fig. 9.2).

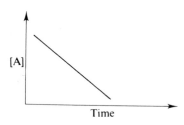

$[A]$

Time

FIG. 9.2. Schematic representation of a zeroth-order process.

---

† We note that the order of the reaction (a, b) may be different from the stoichiometry of the reaction (x, y). The precise relationship between them will depend on the mechanism (i.e. the individual steps) of the reaction.

Zeroth-order processes are not very common. Some examples are found in reactions of gases on metal surfaces, and in enzyme-catalysed reactions at high concentrations of substrate. In these types of processes we have a catalyst which is saturated with the reactant, so that further increases in the concentration of the bulk reactant do not change the reaction rate.

(2) *First-order processes*

$$A \longrightarrow products$$

The rate law for this is:

$$-\frac{d[A]}{dt} = k[A]$$

By integration we obtain

$$\ln[A] = -kt + c$$

where c is the constant of integration.

If $[A]_0$ is the initial concentration of A (i.e. when $t = 0$) then

$$c = \ln[A]_0$$

$$\therefore \quad \boxed{\ln\frac{[A]}{[A_0]} = -kt} \ . \tag{9.1}$$

A plot of $\ln [A]$ vs. t gives a straight line of slope $-k$ (Fig. 9.3).

We have considered the case where we are measuring the disappearance of the reactant $[A]$. However we often study the formation of product $[P]$. In this case the rate law becomes modified as follows.

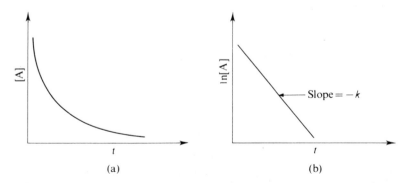

FIG. 9.3. Schematic representation of a first-order process. In practice, plot (*b*) is used to derive the rate constant.

In the process A → P let $[P]_t$ be the concentration of P at time t and $[P]_\infty$ be the final concentration of P. If the reaction is considered irreversible and stoichiometric, $([P]_\infty - [P]_t)$ is the concentration of $[A]$ remaining at time t.
Then

$$\frac{d[P]}{dt} = -\frac{d[A]}{dt}$$

$$= k[A]_t = k([P]_\infty - [P]_t)$$

$$\therefore \quad \frac{d[P]}{([P]_\infty - [P]_t)} = k \, dt.$$

∴ by integration

$$\ln([P]_\infty - [P]_t) = -kt + c$$

when $t = 0$, $[P] = 0$ since we start with pure A.

$$\therefore \quad c = \ln[P]_\infty$$

$$\therefore \quad \ln\left(\frac{[P]_\infty}{[P]_\infty - [P]_t}\right) = kt$$

i.e. a plot of

$$\ln\left(\frac{[P]_\infty}{[P]_\infty - [P]_t}\right) \text{ vs. } t$$

gives a straight line of slope k.

The end point of the reaction can be measured directly, or calculated from the time course of the reaction.

A very useful quantity is the *half-time* for the reaction. This is the time taken for $[A]$ to fall to half its initial value. Let this time be denoted by $t_{\frac{1}{2}}$. Then from the eqn (9.1)

$$\ln 2 = kt_{\frac{1}{2}}$$

$$\therefore \quad t_{\frac{1}{2}} = \frac{\ln 2}{k}$$

$$\boxed{t_{\frac{1}{2}} = \frac{0.693}{k}}.$$

Thus for first-order reactions the half-time is independent of the concentration of A. The time taken for the concentration of A to fall from $[A]_0$ to $0.5[A]_0$ is as long as the time taken for it to fall from $0.5[A]_0$ to $0.25[A]_0$, and so on.

Worked example. Radioactivity decays in a first-order manner. The following data were obtained for the decay of the isotope ^{24}Na.

Time (h)	0	4	8	12	16	20	24
Activity (distintegrations per min)	478	395	329	272	226	187	155

Calculate the rate constant for the decay and half-time (half-life) of the isotope.

Solution. A plot of ln(activity) against time is linear (Fig. 9.4).

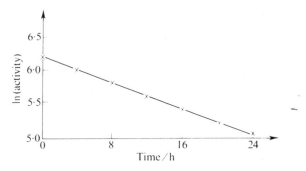

FIG. 9.4. First-order plot for decay of ^{24}Na isotope.

The slope of this graph $(-k)$ equals $-0 \cdot 0469$ h^{-1}. Thus the rate constant (k) is $\underline{0 \cdot 0469\ \text{h}^{-1}}$.

Since

$$kt_{\frac{1}{2}} = \ln 2$$

$$t_{\frac{1}{2}} = \frac{\ln 2}{k}$$

$$= 14 \cdot 8\ \text{h}$$

i.e. the half life = $\underline{14 \cdot 8\ \text{h}}$.

(3) *Second-order processes*

Consider the reaction below

$$A + B \rightarrow \text{products.}$$

Let the initial concentrations of A and B be $[A]_0$ and $[B]_0$, and after time t there are x moles of products formed:

Then

$$\frac{dx}{dt} = k_2[A][B]$$

$$= k_2([A_0] - x)([B]_0 - x).$$

On integration

$$\frac{1}{([A]_0 - [B]_0)} \cdot \ln \frac{[B]_0([A]_0 - x)}{[A]_0([B]_0 - x)} = k_2 t. \qquad (9.2)$$

So a plot of

$$\ln \frac{[B]_0([A]_0 - x)}{[A]_0([B]_0 - x)} \text{ vs. } t$$

gives a straight line of slope $k_2([A]_0 - [B]_0)$.

If the initial concentrations of A and B are equal we cannot use eqn (9.2). In this case

$$\frac{dx}{dt} = k_2([A]_0 - x)^2 \qquad (\text{since } [A]_0 = [B]_0).$$

On integration

$$\frac{1}{([A]_0 - x)} = k_2 t + c.$$

Since $x = 0$ when $t = 0$, then

$$c = \frac{1}{[A]_0}.$$

$$\therefore \quad \frac{x}{[A]_0([A]_0 - x)} = k_2 t.$$

The half-time for the reaction is found by putting $x = \frac{1}{2}[A]_0$.

$$\boxed{t_{\frac{1}{2}} = \frac{1}{k_2[A]_0}}. \qquad (9.3)$$

Thus in second-order reactions the half-time is inversely proportional to the initial concentration of A, i.e. it takes twice as long for the concentration of A to fall from $0.5[A]_0$ to $0.25[A]_0$ as it does to fall from $[A]_0$ to $0.5[A]_0$.

Worked example. The rate constant for the dissociation of NADH from liver alcohol dehydrogenase is $3s^{-1}$. If the dissociation constant of the enzyme–NADH complex is $0.25 \times 10^{-6} \, (\text{mol dm}^{-3})$ what is the rate constant of the association reaction? What is the half-time of the reaction if the initial concentrations of NADH and enzyme are both 100 μmol dm^{-3}.

Solution. For the reaction

$$NADH + enzyme \underset{k_{-1}}{\overset{k_1}{\rightleftharpoons}} enzyme\text{-}NADH$$

$$K_d = \frac{k_{-1}}{k_1}.$$

Now $K_d = 0.25 \times 10^{-6}$ (mol dm^{-3}), $k_{-1} = 3$ s^{-1}.

$$\therefore \quad k_1 = 1.2 \times 10^7 \text{ (mol dm}^{-3})^{-1} \text{ s}^{-1}.$$

The half time for the reaction of NADH and enzyme (second order overall) can be calculated from the formula (eqn (9.3))

$$t_{\frac{1}{2}} = \frac{1}{k[A]_0}$$

where $[A]_0$ is the initial concentration of the reactants, and k is the rate constant.

In this case

$$t_{\frac{1}{2}} = \frac{1}{1.2 \times 10^7 \times 10^{-4}} \text{ s}$$

$$t_{\frac{1}{2}} = 8.3 \times 10^{-4} \text{ s.}$$

Note that the rate of the back reaction is negligible.

(4) *Pseudo-first-order processes*

A reaction of the type

$$A + B \rightarrow products$$

is often of second order overall, with a rate law $-d[A]/dt = k_2[A][B]$. If, in such a case, the concentration of one of the reactants (say B) is much greater than that of the other, then it is clear that during the reaction $[B]$ remains essentially constant. The rate law now becomes

$$-\frac{d[A]}{dt} = k'[A]$$

where k' is a new constant, known as the *pseudo-first-order rate constant*. (Note that $k' = k_2[B]$.)

That is, the process is now *apparently* first order in A (and zeroth order in B). Pseudo-first-order conditions provide a special and useful way of measuring the rate constant of a second-order reaction, since if k' and $[B]$

are known, k_2 can be calculated. In practice, it is found that if [B] is in a 20 fold or greater excess over [A], then pseudo-first-order kinetics will be observed. These conditions often apply in biochemical studies, as for instance when a large concentration of some reagent is used to modify the functional groups of a macromolecule (which is present at fairly low concentrations).

Worked example

Sheep liver pyruvate carboxylase at a concentration of 1 μmol dm^{-3} was modified with cyanate, and the extent of modification was followed by measuring the loss of the acetyl-CoA-dependent activity. From the data given below, determine the pseudo-first-order rate constants (k') and the second-order rate constant (k_2) for the modification.

	[Cyanate] (mmol dm^{-3})			
Time (min)	50	100	200	300
	(% initial activity)			
0	100	100	100	100
10	97	94	89	84
20	94	89	79	71
40	89	79	63	50
60	84	71	50	35
80	79	63	40	25
100	75	56	31	18
150	64	42	18	7·5
200	56	31	10	3

Solution

Because the cyanate concentration is many orders of magnitude greater than the enzyme concentration, then clearly the cyanate concentration remains effectively constant during the modification.

Thus, the rate expression for the modification is:

$$\frac{dA}{dt} = -k'A$$

where A is the enzyme concentration, and k' is the pseudo-first-order rate constant. (Note that $k' = k_2$ [cyanate].)

This expression can be integrated to give

$$\ln(A/A_{initial}) = -k't.$$

Thus k' may be obtained by plotting $\ln(A/A_{initial})$ against time (Fig. 9.5):

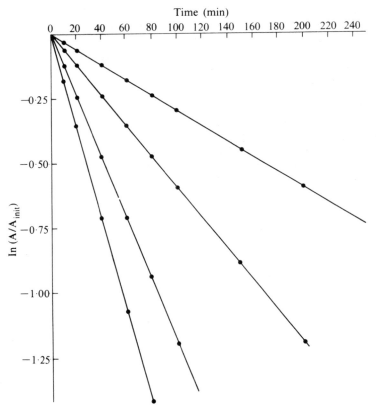

FIG. 9.5. First-order plot to determine the pseudo-first-order rate constant.

k' values may be obtained from this graph either from the slope or by noting that:

$$k' = \frac{0 \cdot 693}{t_{\frac{1}{2}}}$$

(cf. p. 136), where $t_{\frac{1}{2}}$ is the half-time of the reaction.

In this way, the following k' values may be obtained:

[Cyanate] (mmol dm^{-3})	$t_{\frac{1}{2}}$ (min)	k' (min^{-1})
50	236	$2 \cdot 94 \times 10^{-3}$
100	120	$5 \cdot 78 \times 10^{-3}$
200	60	$1 \cdot 16 \times 10^{-2}$
300	40	$1 \cdot 73 \times 10^{-2}$

Now, the second-order rate constant (k_2) may be obtained by noting that:

$$k' = k_2 \text{ [cyanate].}$$

Hence, k_2 is the slope of the line obtained by plotting k' against [cyanate]. Hence

$$k_2 = 5 \cdot 75 \times 10^{-5} \ (\text{mmol dm}^{-3})^{-1} \ \text{min}^{-1}$$

or

$$= \underline{5 \cdot 75 \times 10^{-2} \ (\text{mol dm}^{-3})^{-1} \ \text{min}^{-1}}$$

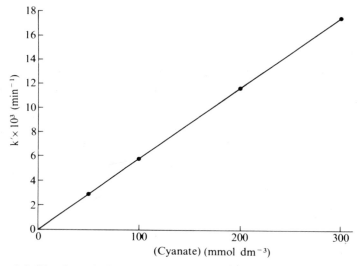

FIG. 9.6. Plot of pseudo-first-order rate constant vs. [cyanate] in worked example to determine second-order rate constant.

(5) *Third-order and higher reactions*

These reactions are extremely rare and are generally of the form

$$2A + B \ \rightarrow \ \text{products.}$$

An example would be the combination of two atoms or radicals which occurs in the presence of a third species.

The treatment of such reactions becomes rather cumbersome and is dealt with in more specialized textbooks.

Determination of the order of a reaction

The order of a reaction can be determined by comparing the experimental data on the concentrations of reactants or products as a function of time with the integrated forms of the various rate laws discussed above. It is worthwhile mentioning that data over as large a percentage of the total reaction as possible should be used, in order to decide the order unequivocally. In the initial phase of a reaction it is difficult to distinguish, for example, between first- and second-order reactions.

Often the determination of reaction orders can be simplified by use of the method of half-times.† It can be shown (see Appendix 3) that if a reaction is of nth order, then

$$t_{\frac{1}{2}} \propto \frac{1}{[A]_0^{n-1}}.$$

This expression will also apply for those reactions involving more than one reactant, provided that the initial concentrations of the reactants are in their stoichiometric proportions.

Thus we can look at the successive half-times for a reaction. In a first-order reaction they are the same; in a second-order reaction they increase in a geometric progression, as is shown in Fig. 9.7.

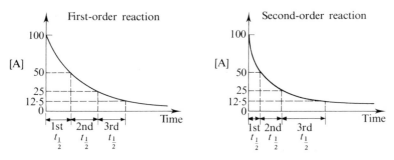

FIG. 9.7. Time courses of a first-order and a second-order reaction, illustrating the method of half-times.

The initial concentration of $[A]$ is taken to be 100 in both cases. In the first-order reaction it is clear that the successive half-times are the same. In the second-order reaction they are in the proportion $1:2:4$, etc.

---

† The same equation applies for *any* fractional time, e.g. $t_{0.8}$, $t_{0.6}$, or $t_{0.25}$, etc.

The overall order of the reaction can be determined by this method. There are two main ways to determine the order with respect to any one reactant (e.g. A). The first method involves keeping all the other reactants in a large excess, and then studying the reaction rate as a function of A. This procedure is then varied so as to examine each of the reactants in turn.

In the second method, studies of the initial rate of the reaction are made.

$$\text{Initial rate} = -\frac{d[A]}{dt} = k[A]_0^n \cdot [B]_0^m \cdot [C]_0^p$$

where $[A]_0$, $[B]_0$, and $[C]_0$ are the initial concentrations of A, B and C; and n, m, and p are the orders with respect to these reactants.

The concentrations of B and C are kept constant, and $[A]_0$ is varied. Analysis of the variation of initial rate with $[A]_0$ will then given the order with respect to A. The procedure is again varied to determine the orders with respect to B and C in turn.

Worked examples

(1) The rate of formation of a product P in a reaction was studied.

Concentration of P								
formed/μmol dm^{-3}	0	34·3	56·8	71·6	81·7	88·0	92·1	100
Time/min	0	2·5	5·0	7·5	10·0	12·5	15·0	∞

What is the order of the reaction?

Solution

If the concentration of P is plotted against time (Fig. 9.8), we can deduce successive half times for the reaction of 4·25 min, 4 min, and 4·25 min.

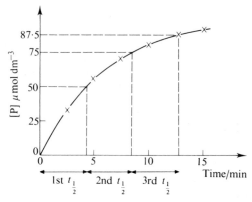

FIG. 9.8. Data of 'Worked example' plotted to illustrate the method of half-times.

Since these are effectively constant the reaction is first order. The rate constant k can be evaluated from $kt_{\frac{1}{2}} = \ln 2$, i.e.

$$k = \frac{0 \cdot 693}{4 \cdot 25} \, \text{min}^{-1}$$

$$= \underline{0 \cdot 163} \, \text{min}^{-1}.$$

k could also be evaluated from the appropriate first-order plot.

(2) The important biochemical intermediate, acetyl CoA, may be prepared by reacting CoASH with acetyl chloride, the reaction being followed by measuring the CoASH concentration at various times. The following data were obtained when the initial concentrations of both reagents were 10 mmol dm^{-3}

Time (min)	0	1	1·5	2	2·75	4·41	5
[Acetyl CoA] (mmol dm^{-3})	0	3·3	4·2	4·8	5·8	6·9	7·0

Determine the order of the reaction, and evaluate the relevant rate constant.

Solution

First, we shall assume that the reaction is a first order process. If this is the case, then a graph of $\ln(\text{CoA}/\text{CoA}_{\text{initial}})$ vs. time should give a straight line. From the data, we have:

Time (min)	0	1	1·5	2	2·75	4·41	5
[CoA] (mmol dm^{-3})	10	6·7	5·8	5·2	4·2	3·1	3·0
$\text{CoA}/\text{CoA}_{\text{init}}$	1	0·67	0·58	0·52	0·42	0·31	0·30
$\ln(\text{CoA}/\text{CoA}_{\text{init}})$	0	−0·40	−0·54	−0·65	−0·87	−1·17	−1·20

A straight line is not obtained, and hence it may be concluded that the reaction is not first order (see Fig. 9.9).

We shall now assume that the reaction is second order. Because the initial [CoA] and [acetyl chloride] are equal, we can use the equation:

$$\frac{[\text{acetyl CoA}]}{[\text{CoA}]_{\text{init}}([\text{CoA}_{\text{init}} - \text{acetyl CoA}])} = k_2 t$$

(cf. p. 138).
From the data we obtain:

Time (min)	0	1	1·5	2	2·75	4·41	5
$\dfrac{[\text{acetyl CoA}]}{[\text{CoA}]_{\text{init}}[\text{CoA}_{\text{init}} - \text{acetyl CoA}]}$	0	0·049	0·072	0·092	0·138	0·223	0·233

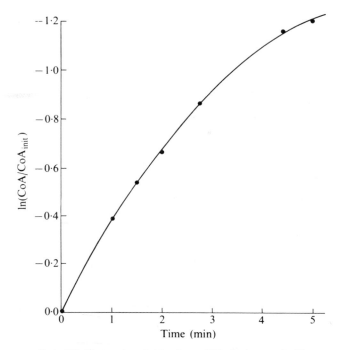

FIG. 9.9 First-order plot for data of Worked example (2).

In this case, a good straight line is obtained, which indicates that the reaction is second order (see Fig. 9.10). From the slope of the line, the second order rate constant is $4 \cdot 83 \times 10^{-2}$ $(\text{mmol dm}^{-3})^{-1} \text{min}^{-1}$.

A note on units

It is clear that the first-order rate constants contain no unit of concentration. The units are simply $(\text{time})^{-1}$ (e.g. s^{-1}, min^{-1}, h^{-1}, etc.)

Second-order rate constants contain concentration and time units. From the rate law, the units are seen to be $(\text{concentration})^{-1} (\text{time})^{-1}$, (e.g. $(\text{mol dm}^{-3})^{-1} \text{s}^{-1}$, $(\text{mol dm}^{-3})^{-1} \text{h}^{-1}$ etc.).

Pseudo-first-order rate constants are clearly in units of $(\text{time})^{-1}$. Since they 'include' the concentration of the reactant which is in excess, they can be corrected so as to give true second-order rate constants by dividing by this concentration. Thus if the concentration of B is 1 mmol dm^{-3} (B is in excess) and the pseudo-first-order rate constant obtained is 10^{-2}min^{-1}, then the true second order rate constant is $(10^{-2}/10^{-3}) (\text{mol dm}^{-3})^{-1} \text{min}^{-1}$; i.e. $10 (\text{mol dm}^{-3})^{-1} \text{min}^{-1}$.

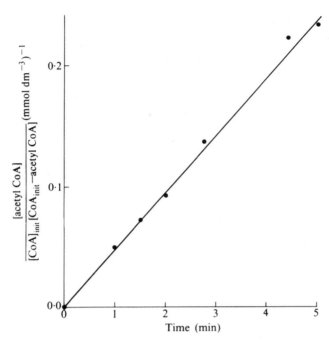

FIG. 9.10. Second-order plot for data of Worked example (2).

The kinetics of some other types of processes

In this section we will examine briefly various other types.

(1) *Parallel reactions*

A compound A could give rise to two distinct products:

$$A \xrightarrow{k_1} B$$
$$A \xrightarrow{k_2} C$$

with respective rate constants k_1 and k_2. If both reactions are first-order with respect to A, we have

$$-\frac{d[A]}{dt} = (k_1 + k_2)[A]$$

$$\frac{d[B]}{dt} = k_1[A]$$

$$\frac{d[C]}{dt} = k_2[A].$$

Thus the disappearance of A is still first order, but the overall rate constant is now the sum of the rate constants for the individual processes.

Examples of this type of reaction are known in organic chemistry, e.g. at 1000 K acetic acid gives CH_4 and CO_2 in one reaction and ketene and H_2O in a parallel reaction. Many examples can be found in which a biochemical compound is converted into two other compounds, e.g. AMP can be converted to ATP or to IMP. However these are generally enzyme catalysed reactions and the kinetics of these reactions are more complicated than those discussed here. Other examples can be found in isotope work. When an isotopically-labelled compound is injected into an animal, the amount of isotope remaining generally decreases in a first-order fashion with time. There are two processes responsible, namely the radioactive decay of the isotope and the loss of the compound from animal by excretion. Both of these, then, are first order.

(2) *Reversible reactions*

If the rate of the back reaction becomes significant, then the rate laws are more complex.

Consider the reaction:

$$A \underset{k_{-1}}{\overset{k_1}{\rightleftharpoons}} B$$

where k_1 and k_{-1} are the rate constants for the forward and back reactions. If x moles of A have been converted to B at time t then:

$$\frac{dx}{dt} = k_1(A_0 - x) - k_{-1}x$$

where A_0 is the initial concentration of A, and the initial concentration of B is assumed to be zero.

Because the reaction is reversible, at some time it will reach equilibrium, with no net formation of A or B, i.e. the reaction rates in both directions will be equal.

Hence, if x_e is the equilibrium concentration of B

$$k_1(A_0 - x_e) = k_{-1}x_e \tag{9.4}$$

$$\therefore \quad k_{-1} = \frac{k_1(A_0 - x_e)}{x_e}.$$

Hence

$$\frac{dx}{dt} = k_1(A_0 - x) - \frac{k_1(A_0 - x_e)}{x_e} \cdot x$$

i.e.

$$\frac{dx}{dt} = \frac{k_1A_0}{x_e}(x_e - x).$$

As A_0, k_1, and x_e are constants, this expression may be integrated noting that $x = 0$ when $t = 0$, to give

$$\ln \frac{x_e}{x_e - x} = \frac{k_1 A_0}{x_e} \cdot t$$

This expression can be expressed in an alternative form, using the equilibrium relation (9.4) to give

$$\ln \frac{x_e}{x_e - x} = (k_1 + k_{-1})t \tag{9.5}$$

Note that the rate constant for the approach to equilibrium is equal to the sum of the first-order rate constants k_1 and k_{-1}.

A fuller treatment of this important case can be found in the text by Fersht. (See reading list.)

Eqn (9.5) is sometimes written in the form:

$$\ln \frac{\Delta C_0}{\Delta C_e} = (k_1 + k_{-1})t$$

where ΔC_0 is the difference between the initial and equilibrium concentrations of A (equal to x_e if there is no B present at $t = 0$) and ΔC_e is the difference between the concentration of A at time t, and at equilibrium. In this form, the equation will apply regardless of whether there is any B present at the start.

This equation is the basis of *relaxation techniques* for the kinetic study of reversible reactions. If the reaction is at equilibrium and the equilibrium is suddenly altered (e.g. by altering the temperature or pressure) then the equilibrium will 'relax' to a new state, with a time course defined by the above equation, if we have opposed first-order reactions. This technique has great advantages for studying *rapid reactions* (such as the reversible binding of a substrate to an enzyme) because it avoids the need to start the reaction by mixing the reactants (which takes a finite time of the order of 1 ms). The temperature can be changed very rapidly by 5 or 10 K in about 10 μs by discharging a condenser between electrodes in the solution. By suitable measuring devices the change in concentration in A and B resulting from the change in the position of equilibrium† can be assessed. This is the *temperature jump* method for studying rapid reactions in solution—both inorganic (e.g. $H^+ + OH^- \rightarrow H_2O$, $k = 1 \cdot 5 \times 10^{11}$ (mol dm^{-3})$^{-1}$ s^{-1}) and also elementary steps involved in enzyme catalysis (see Chapter 10). Even if there are second-order reactions involved in the relaxation of an equilibrium, the time course will follow a first-order course provided that the perturbation of the equilibrium is not too great. This greatly simplifies the mathematical analysis of more complex reactions.

---

† The position of equilibrium will only be changed by altering the temperature or pressure if for the reaction ΔH° or ΔV° (the change in volume) are non zero respectively.

Worked example

One interesting biochemical system to which reversible first order equations apply is the carbonic acid system:

$$CO_2 + H_2O \underset{k_{-1}}{\overset{K_a}{\rightleftharpoons}} H_2CO_3 \underset{\text{very fast}}{\overset{K_a}{\rightleftharpoons}} H^+ + HCO_3^- \qquad (9.6)$$

The forward reaction rate r_f is given by the expression

$$r_f = k_1[CO_2][H_2O]$$

since the $[H_2O]$ is essentially constant, this can be written as:

$$r_f = k_f[CO_2].$$

The reverse reaction rate r_b is given by the expression:

$$r_b = k_{-1}[H_2CO_3]$$

Now, the dissociation of H_2CO_3 is very fast, and will always be at equilibrium.

Hence

$$[H_2CO_3] = \frac{[H^+][HCO_3^-]}{K_a}.$$

Therefore

$$r_b = \frac{k_{-1}[H^+][HCO_3^-]}{K_a}.$$

This means that eqn (9.6) reduces to

$$CO_2 \underset{k_b}{\overset{k_f}{\rightleftharpoons}} HCO_3^-$$

where

$$k_b = \frac{k_{-1}[H^+]}{K_a}.$$

Therefore, the rate at which CO_2 dissolves in a buffer solution of fixed pH can be simplified to two opposing first-order reactions.

(3) *Consecutive reactions*

An important class of reactions are those in which the product of a reaction is subsequently converted to a second product.

Consider the process

$$A \xrightarrow{k_1} B \xrightarrow{k_2} C.$$

The rate equations can be written down easily:

$$-\frac{d[A]}{dt} = k_1[A]$$

$$\frac{d[B]}{dt} = k_1[A] - k_2[B]$$

$$\frac{d[C]}{dt} = k_2[B].$$

The solution of these equations is rather long and we shall just note that the concentrations of A, B, and C at various times are approximately as shown in Fig. 9.11 (it is assumed that the rate constants k_1 and k_2 are of similar magnitudes).

In this example it is seen that the concentration of the *short lived* intermediate B rises at first, reaches a maximum value, and then falls slowly to zero. This type of result forms the basis of the so-called '*steady-state approximation*', which states that the concentration of any intermediate reactive species can be considered to remain constant (i.e. the rate of its formation equals the rate of its breakdown). The time during which this is true is known as the '*steady state*' period. As is clear from the above diagram, there is a time in which the concentration of B rises rapidly; this is known as the *pre-steady state period. The subsequent rate of change of* [B] *is very small compared with the rates of change of* [A] *and* [C] *and under these conditions the steady state approximation is valid.*†

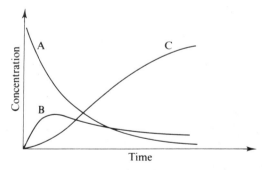

FIG. 9.11. The concentrations of A, B, and C as a function of time for the consecutive reaction scheme.

---

† It is only under these conditions that the steady state approximation is valid. It is *not* valid when the rate of change of the concentration of the intermediate is comparable with the rates of change of the concentrations of either reactants or products.

If the concentrations of B remains constant, this would be expressed mathematically as

$$\boxed{\frac{d[B]}{dt} = 0}.$$

The importance of the steady state approximation is that it greatly simplifies the solution of the rate equations in complex consecutive reaction pathways.

Consider the interaction of an enzyme E with its substrate S to form an enzyme-substrate complex, which can then break down to give the product P and regenerate E.

$$E + S \underset{k_{-1}}{\overset{k_1}{\rightleftharpoons}} ES \overset{k_2}{\rightarrow} E + P$$

According to the steady state approximation the rate of production of ES (i.e. $k_1[E][S]$) equals the rate of its breakdown (i.e. $k_{-1}[ES] + k_2[ES]$),

$$k_1[E][S] = k_{-1}[ES] + k_2[ES]$$

$$[ES] = \frac{k_1[E][S]}{k_{-1} + k_2}.$$

The rate of product formation† is then given by

$$\frac{d[P]}{dt} = k_2[ES]$$

and by substituting for [ES] from the previous equation, the rate of product formation can be written down. It should be pointed out that this equation is in terms of the concentration of free enzyme [E] and not in terms of the total concentration of enzyme. This modification is referred to in Chapter 10 on enzyme kinetics.

The steady state approximation thus allows us to set up equations for each reactive intermediate in a reaction. These equations can then be solved to find the (steady state) concentrations of these intermediates. Derivation of the overall rate of the reaction is then a relatively simple task.

Having dealt with the rate laws for various types of processes we can now proceed to examine the effects of some of the variables which can give information on the mechanism of a reaction. The effect of temperature will be considered first.

---

† Provided [ES] is constant and small compared with S, the rate of product formation equals the rate of disappearance of substrate.

Effect of temperature on the rate of a reaction

It has long been observed that the rate of most reactions increases dramatically with temperature. The equation proposed by Arrhenius was of the form:†

$$\frac{d \ln k}{dT} = E_a / RT^2$$

where k is the rate constant, E_a is the 'activation energy' for the reaction, and R is the gas constant.

If E_a is assumed to be independent of temperature, this equation can be integrated to give

$$\ln \left(\frac{k_{T_1}}{k_{T_2}} \right) = -\frac{E_a}{R} \left(\frac{1}{T_1} - \frac{1}{T_2} \right)$$

or

$$k = A \exp(-E_a / RT) \; ; \tag{9.7}$$

A is known as the *pre-exponential* factor.

A plot of $\ln k$ vs. $1/T$, which is known as the *Arrhenius plot*, gives a straight line, from the slope of which E_a can be obtained. The value of A is obtained by extrapolation of the graph to $1/T = 0$.

Worked example

A reaction rate doubles on going from 293 K to 303 K. What is the activation energy of the reaction? ($R = 8.31 \text{ J K}^{-1} \text{ mol}^{-1}$.)

Solution

In eqn (9.7)

$$\ln \frac{k_{T_1}}{k_{T_2}} = -\frac{E_a}{R} \left(\frac{1}{T_1} - \frac{1}{T_2} \right)$$

---

† We might note the 'analogy' between the Arrhenius equation and the van't Hoff isochore, eqn (3.7)

$$\frac{d \ln K}{dT} = \frac{\Delta H^\circ}{RT^2},$$

and remember that

$$K = \frac{k_1}{k_{-1}}.$$

we put

$$\frac{k_{T_1}}{k_{T_2}} = \tfrac{1}{2}, \qquad T_1 = 293 \text{ K}, \qquad T_2 = 303 \text{ K}$$

from which

$$E_a = 51 \cdot 4 \text{ kJ mol}^{-1}.$$

This result provides a handy 'rule of thumb'. If a reaction rate doubles for an increase of 10 K around room temperature, the activation energy is about $51 \cdot 0$ kJ mol^{-1}. A reaction whose E_a is less than $51 \cdot 0$ kJ mol^{-1} is less sensitive to temperature. Compare the example in Chapter 3 (p. 25) dealing with the temperature variation of the equilibrium constant of a reaction.

Significance of the parameters in the Arrhenius equation

Reactions are presumed to proceed via collisions between molecules. In gaseous systems, kinetic theory can be used to calculate the number of such collisions occurring under a given set of conditions. Not all collisions, however, will lead to products since only a certain fraction of the molecules will possess the necessary 'activation energy' so that collision can lead to reaction. We might then expect that the rate of the reaction would be given by

$$\text{Rate} = Z_{12} \exp(-E_a/RT)$$

where Z_{12} is the collision number (in suitable units) and $\exp(-E_a/RT)$ is the Boltzmann factor giving the fraction of the molecules which possess the necessary activation energy E_a. The thermal energy available is RT at T K.

The Z_{12} (collision) term would appear to correspond to the pre-exponential factor A in the Arrhenius equation. However it was found in practice that for most second-order gas reactions the calculated value of Z_{12} is very different from the observed value of A. An arbitrary 'steric factor' p was then included in the Arrhenius expression:

$$k = pA \exp(-E_a/RT).$$

A better insight into the significance of these parameters is afforded by a brief account of the *transition state* theory of reaction rates.

Transition state theory

The collision theory assumes that molecules are hard spheres and only considers whether the molecules have enough energy to overcome the energy barrier. It takes no account of the nature of the reacting species. The transition state theory concentrates on the configuration of the reactants just as they are about to pass over an energy barrier and become products. This configuration is thought of as being at the highest energy point of the energy profile describing the reaction (Fig. 9.12). It is called the transition state or activated complex.

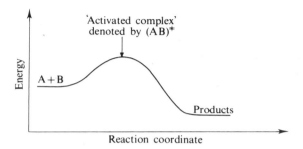

FIG. 9.12. The energy profile of a reaction showing the activated complex.

In the most easily understood form of the theory, it is assumed† that an equilibrium exists between the reactants and the activated complex, characterized by an equilibrium constant, $K^{\ddagger}$

$$K^{\ddagger} = \frac{[AB^{\ddagger}]}{[A][B]}. \tag{9.8}$$

For this equilibrium we can assign values of $\Delta G^{\circ\ddagger}$, $\Delta H^{\circ\ddagger}$, and $\Delta S^{\circ\ddagger}$. (These are known as the *free energy, enthalpy,* and *entropy of activation,* respectively.)

The great conceptual advantage of this treatment is that it allows the application to reaction rates of all the basic thermodynamic considerations of reaction equilibria—thus forming the link between thermodynamics and kinetics. The values of $\Delta H^{\circ\ddagger}$ and $\Delta S^{\circ\ddagger}$ will also reflect the nature of the reacting species.

The rate at which the activated complexes are converted to products is then assumed to be slow enough so as not to disturb this equilibrium significantly. From the theory, the rate of such conversion (by passage over the energy barrier) is kT/h, where k is the Boltzmann constant and h is Planck's constant.

The overall rate is then given by:

$$-\frac{d[A]}{dt} = [AB^{\ddagger}] \cdot \frac{kT}{h}$$

---

† Strictly speaking this treatment of transition state theory is not rigorous, but in practice this introduces only a small error.

and since

$$-\frac{d[A]}{dt} = k'[A][B]$$

the rate constant k' is given by

$$k' = \frac{kT}{h} \cdot \frac{[AB^{\ddagger}]}{[A][B]}$$

$$= \frac{kT}{h} \cdot K^{\ddagger} \qquad \text{(from eqn (9.8))}$$

$$= \frac{kT}{h} \exp\left(-\frac{\Delta G^{\circ\ddagger}}{RT}\right)$$

$$= \frac{kT}{h} \exp\left(+\frac{\Delta S^{\circ\ddagger}}{R}\right) \cdot \exp\left(-\frac{\Delta H^{\circ\ddagger}}{RT}\right).$$

which is of the form

$$\boxed{k' = A \exp(-E_a/RT)} \ .$$

The $\Delta H^{\circ\ddagger}$ is more accurately related to the Arrhenius activation energy (E_a) (which is an internal energy change) by the equation

$$\Delta H^{\circ\ddagger} = E_a + \Delta(PV^{\circ\ddagger})$$

The pre-exponential factor in the Arrhenius expression† involves the entropy of activation $\Delta S^{\circ\ddagger}$. The values of $\Delta H^{\circ\ddagger}$ and $\Delta S^{\circ\ddagger}$ allow some insight into the factors governing the rates of reactions.

Meaning of $\Delta H^{\circ\ddagger}$ and $\Delta S^{\circ\ddagger}$

In general, $\Delta H^{\circ\ddagger}$ is a measure of the energy barrier which must be overcome by reacting molecules and is related to the strength of the bonds which are broken and made in the formation of the transition state from the reactants.

$\Delta S^{\circ\ddagger}$ is related to how many molecules with the appropriate energy can actually react. The value of $\Delta S^{\circ\ddagger}$ includes steric and orientation requirements and also solvent effects, and provides a better insight into the roles of the reactants than the much less definite probability factor of the collision theory.

For example, *bimolecular* reactions would be expected to have more

---

† It should be noted that the frequency factor kT/h is also temperature dependent, but of course the variation of this term with temperature is generally much less than that of the exponential term, $\exp(-\Delta H^{\circ\ddagger}/RT)$.

negative $\Delta S^{o\ddagger}$ values than monomolecular reactions since bringing two molecules together results in more ordering. *Denaturation*, by contrast has large positive $\Delta S^{o\ddagger}$ values since the organized structure of a macromolecule becomes disordered. In the case of boiling an egg for instance the value of $\Delta S^{o\ddagger}$ is the dominant factor in the rate equation.

Worked example

Comment on the following relative rates of intramolecular hydrolysis of esters by the carboxylate anion.

1	20	230	10 000

Solution

Transition states of reactions have strict requirements of orientation of the participating species. As the initial orientation of the COO^- and $COOR$ groupings approach that in the transition state, the molecules require less ordering, $\Delta S^{o\ddagger}$ decreases, and the rate increases. However, many factors determine $\Delta S^{o\ddagger}$ and so care should be used in any quantitative interpretation of it. For instance the reaction between ions in aqueous solution

$$Ce^{4+} + EDTA^{4-} \rightarrow (CeEDTA)$$

has a large positive value of $\Delta S^{o\ddagger}$. This is because both reactants are highly solvated and the activated complex has a much lower charge and is therefore much less solvated. The positive entropy change associated with the release of water easily overcomes the negative value associated with the reaction of two species to give a complex.

We might also see how the well known S_N1 and S_N2 reactions in organic chemistry would have different values of $\Delta S^{o\ddagger}$.

$$S_N1 \quad AB \rightarrow A^+ \ldots B^-$$

(separation of ions in transition state, hence a positive $\Delta S^{o\ddagger}$ expected)

$$S_N2 \quad AB + C^- \rightarrow (C \ldots A \ldots B)^-$$

(decrease in the number of species in the transition state, hence a negative $\Delta S^{o\ddagger}$ expected).

This shows how the useful concept of the 'entropy of activation' replaces the much less definite 'steric factor' in the old collision theory of reaction rates.

The second variable to be considered in this section is the effect of pH.

The effect of pH on the rate of a reaction

We can divide this section into two principal categories.

(1) Those reactions in which the ionization of groups directly affects the reactivity of the reactants.
(2) Reactions subject to acid or base catalysis.

(1) *The effect of ionizing groups*

Many examples can be found in which, say, a basic form of a compound is a powerful nucleophile, whereas its conjugate acid is unreactive, e.g.

$$\underset{\text{unreactive}}{RNH_3^+} + H_2O \rightleftharpoons \underset{\text{reactive}}{RNH_2} + H_3O^+.$$

Clearly as the pH is raised, $[RNH_2]$ is increased and hence the reaction rate increases. Now

$$K_a = \frac{[RNH_2][H_3O^+]}{[RNH_3^+]}$$

$$\therefore \quad [RNH_2] = \frac{[RNH_3^+] \cdot K_a}{[H_3O^+]}.$$

The fraction in the unprotonated form (F) is:

$$F = \frac{[RNH_2]}{[RNH_2] + [RNH_3^+]}$$

$$= \frac{K_a/[H_3O^+]}{(K_a/[H_3O^+]) + 1}$$

$$= \frac{K_a}{K_a + [H_3O^+]}.$$

The rate constant k is given by

$$k = k_0 F$$

where k_0 is the intrinsic rate constant for the unprotonated form;

$$\therefore \quad k = k_0 \frac{K_a}{K_a + [H_3O^+]}.$$

Now if $[H_3O^+] \gg K_a$ (i.e. pH $<$ pK_a) then

$$k \approx k_0 \frac{K_a}{[H_3O^+]}$$

$$\therefore \quad \log_{10} k = \log_{10} k_0 - pK_a + pH$$

so that a plot of $\log_{10} k$ vs. pH will be linear with a slope of unity. (This is observed provided $pH \leqslant pK_a - 1.5$.) It should be noted that if such behaviour is observed it can be concluded that the unprotonated form is the reactive species, and that the conjugate acid is unreactive.

Note that when $[H_3O^+] \ll K_a$ (i.e. $pH > pK_a$)

$$k = k_0.$$

The overall plot of $\log_{10} k$ vs. pH is as illustrated in Fig. 9.13.

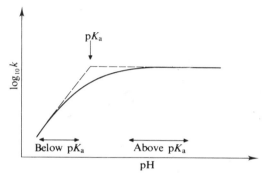

FIG. 9.13. Graph of $\log_{10}$ (rate constant) vs. pH. The solid line is the observed behaviour, the dotted lines are extrapolations of the linear parts.

From the equation it is easily seen that at the intersection point $[H_3O^+] = K_a$, i.e. this method can be used to find the pK_a of an ionizing group.

(2) *Acid–base catalysis*

This is a very large subject and will only be treated in outline here.

Acid–base catalysis can be subdivided into general or specific catalysis. For instance *general acid catalysis* refers to contributions from all the acidic species present in the solution (e.g. H^+, CH_3CO_2H, H_2O etc.). The overall rate equation thus includes terms representing each of these species. In *specific acid catalysis* we consider only the term in H^+ (H_3O^+ in aqueous solution), since the undissociated acids do not contribute. An experimental distinction between the two types could be made by varying the concentration of undissociated acid at constant pH (i.e. by varying the concentration of the conjugate base). As the concentration of the undissociated acid is increased, the rate of a reaction subject to general acid catalysis will rise, whereas the reaction subject to specific acid catalysis will be unaffected.

In order to understand more clearly why some reactions are subject to *specific* and others to *general* acid–base catalysis, we must examine the various steps involved.

We can consider the general scheme

$$A + HX \underset{k_{-1}}{\overset{k_1}{\rightleftharpoons}} AH^+ + X^- \tag{9.9}$$

$$AH^+ + B \xrightarrow{k_2} \text{products.} \tag{9.10}$$

Rate of reaction $= k_2[AH^+][B]$.

Treating AH^+ as a reactive intermediate, and using the steady state approximation, then

$$\frac{d[AH^+]}{dt} = 0 = k_1[A][HX] - k_{-1}[AH^+][X^-] - k_2[AH^+][B]$$

$$\therefore \boxed{[AH^+] = k_1[A][HX]/(k_{-1}[X^-] + k_2[B])} \ . \tag{9.11}$$

We can consider *two limiting cases* of eqn (9.11). In the first, $k_{-1}[X] \gg k_2[B]$—this corresponds to the equilibrium in eqn (9.9), being rapidly established and then being disturbed only slightly by a subsequent slow step (9.10). We obtain

$$[AH^+] = k_1[A][HX]/k_{-1}[X^-]$$

substituting for $[HX]/[X^-]$ using the expression

$$K_{HX} = [HX]/[H^+][X^-]$$

then

$$\boxed{\text{Rate} = \frac{k_1}{k_{-1}} \cdot K_{HX}[H^+][A]} \ .$$

The rate law is of the form corresponding to *specific acid catalysis*. The appropriate reaction profile is shown in Fig. 9.14. The second limiting case of eqn (9.11) is when $k_2[B] \gg k_{-1}(X^-)$—i.e. the rate is determined by the forward reaction in eqn (9.9) and we obtain

$$\text{Rate} = k_2[B][AH^+] = \boxed{k_1[A][HX]} \ .$$

If other acids, e.g. HX', HX'' are present, the rate would be:

$$\text{Rate} = k_1[A][HX] + k_1'[A][HX'] + k_1''[A][HX''] + \ldots,$$

i.e. this is a rate equation corresponding to *general acid catalysis*. The energy profile of the reaction is shown in Fig. 9.15.

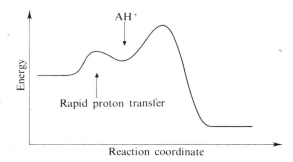

FIG. 9.14. Energy profile for a reaction in which proton transfer occurs in a rapidly established equilibrium followed by a slow step.

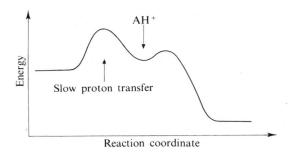

FIG. 9.15. Energy profile for a reaction in which proton transfer is the slow step.

An example of the difference between *general* and *specific* catalysis is afforded by the aldol condensation reactions of acetone and acetaldehyde. The reaction of acetone to give diacetone alcohol is subject to *specific base* catalysis, indicating that a rapid proton abstraction occurs before the slow step:

$$CH_3COCH_3 + OH^- \rightleftharpoons \overset{-}{C}H_2COCH_3 + H_2O \quad \text{(fast)}$$

$$\overset{-}{C}H_2COCH_3 + CH_3COCH_3 \xrightarrow{k_2} \text{diacetone alcohol} \quad \text{(slow)}$$

$$\text{Rate} = k_2[CH_2COCH_3][CH_3COCH_3],$$

$$= k_2 K'_{eq}[CH_3COCH_3]^2[OH^-],$$

where

$$K'_{eq} = \frac{[\overset{-}{C}H_2COCH_3]}{[CH_3COCH_3][OH^-]}.$$

The corresponding reaction of acetaldehyde is subject to *general base* catalysis, implying that proton abstraction is the slow step.

$$CH_3CHO + OH^- \xrightarrow{k_1} \bar{C}H_2CHO + H_2O \qquad \text{(slow)},$$

$$\bar{C}H_2CHO + CH_3CHO \xrightarrow{k_2} \text{aldol} \qquad \text{(fast)}.$$

The rate law contains contributions from all the bases (e.g. OH^-, H_2O, or added $CH_3CO_2^-$) in the solution

$$\text{Rate} = k_1[CH_3CHO][OH^-] + k_1'[CH_3CHO][H_2O] + \ldots.$$

Presumably the difference between the two reactions can be attributed to the fact that the carbonyl group in acetone is less reactive than that in acetaldehyde (a result of the deactivating effect of a methyl group), and thus nucleophilic attack on the group, which is the second step of the reaction, is slower in the case of acetone.

The study of the mechanisms of acid–base-catalysed reactions is of considerable value in providing models for enzyme catalysis, where amino acid side chains are often postulated to act as general acids or bases. The following reactions depict imidazole acting as a general base in a model reaction, and part of a proposed mechanism of action of ribonuclease in which histidine acts as a general base (Fig. 9.16).

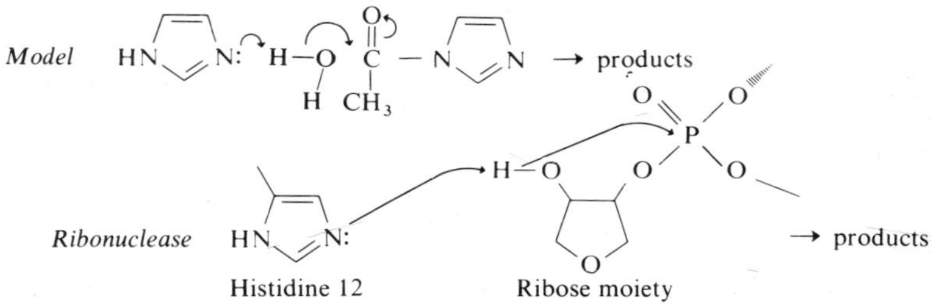

FIG. 9.16. Comparison of mechanism of ribonuclease with that of a model reaction.

The effect of ionic strength on the rate of a reaction

A third parameter which can give some information on the mechanism of the reaction is the *ionic strength* of the solution.

This effect (known as the *salt effect*) on the reaction arises because changes in the ionic strength will change the activity coefficients of the reactants (if they are charged). Thus in the transition state theory expression:

$$k' = \frac{kT}{h} \cdot K^{\ddagger}.$$

We now have to express $K^{\ddagger}$ in terms of activities (rather than concentrations). The relationship between activity coefficients and ionic strength is given by the Debye–Hückel theory (p. 56). For aqueous solutions of low ionic strength (at 298 K), the final equation derived is:

$$\log_{10} \frac{k'}{k_0} = Z_A \cdot Z_B \sqrt{(I)}.$$

Where k' is the rate constant at an ionic strength I.

k_0 is the rate constant at zero ionic strength (obtained by extrapolation).

Z_A, Z_B are the charges (with signs) of the ions taking part in the reaction A + B → products.

The behaviour predicted by this equation has been observed experimentally in many cases (provided that the ionic strength is low enough for the Debye–Hückel theory to be valid). Note that if one of the reactants is uncharged, the product $Z_A \cdot Z_B$ is zero and the rate is independent of the ionic strength.

This effect is known as the *primary salt effect*. A second type of salt effect is observed in acid–base catalysis when changes in the ionic strength can affect the degree of dissociation of weak acids and bases and hence the concentrations of the catalytic species. This indirect effect on the reaction rate is known as the *secondary salt effect*.

A final parameter to be mentioned is the effect of isotopic substitution on reaction rates.

The effect of isotope changes on reaction rates

If an atom at a known position in a molecule is replaced by a different isotope of the atom, then observation of an effect on the rate of reaction of the molecule gives a very good indication that a bond to that atom is being broken in the rate determining step of the reaction.

Consider the replacement of H in a C—H bond by D (deuterium). Because of the heavier mass of D, the zero point energy of the C—D bond is less than that of C—H (Fig. 9.17).

The dissociation energy of the C—D bond is thus approximately $5\cdot0$ kJ mol^{-1} greater than that of the C—H bond. From the $\exp(-E_a/RT)$ term in the Arrhenius expression, this would be expected to decrease the

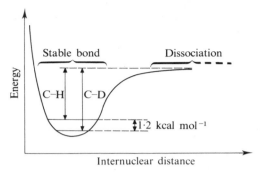

FIG. 9.17. Schematic representation of the energy of C—H and C—D as a function of distance showing the zero-point energies of C—H and C—D.

rate of a reaction whose rate determining step involves breaking the bond by a factor of 7·6 when H is replaced by D. Several 'isotope effects' of this magnitude are known, e.g. in the oxidation of benzaldehyde by $KMnO_4$ at pH = 7, but generally the values for the ratio k_{C-H}/k_{C-D} are rather lower (~2).

The lower value could mean that in the activated complex the C—H bond may not be completely broken, or a new bond from the H to some other atom may be substantially formed. In some reactions (e.g. the nitration of benzene) no isotope effect is observed when H is replaced by D, and thus it is clear that breaking of a C—H bond is not involved in the rate determining step of the reaction.

Thus we see how kinetic investigations (including the form of the rate law, effects of temperature, pH, etc.) can be used to provide an insight into the pathway of a reaction. When this information is combined with, for example, data on the stereochemistry of the reaction, or isolation of intermediates it is often possible to build up a fairly detailed picture of the mechanism. Several notable examples are known in organic chemistry (especially substitution and elimination reactions) and more recently in inorganic systems (e.g. ligand-exchange reactions among transition metal ions).

PROBLEMS

1. Distinguish between the molecularity and order of a chemical reaction.
   An organic acid decarboxylates spontaneously at 298 K and at pH = 6. The evolution of CO_2 from 10 cm^3 of a 10 mmol dm^{-3} solution was measured at various times as follows:

Time/min	25	50	75	100	125	150	200
CO_2 evoloved/cm^3	0·64	1·10	1·45	1·70	1·90	2·05	2·23

Determine the order and rate constant for this reaction. Assume that no CO_2 remains in solution.

2. The compound A can give two alternative products B and C.

$$A \rightarrow B$$

$$A \rightarrow C$$

The first-order rate constants are $0 \cdot 15$ min^{-1} and $0 \cdot 06$ min^{-1} respectively. What is the half-life of A?

If the initial concentration of A is $0 \cdot 1$ mol dm^{-3}, at what time does the concentration of B $= 0 \cdot 05$ mol dm^{-3}?

What is the maximum concentration of B?

3. The following results were obtained for the reaction of 1 mmol dm^{-3} N-acetylcysteine with 1 mmol dm^{-3} iodoacetamide.

N-acetylcysteine concentration/mmol dm^{-3}	0·770	0·580	0·410	0·315	0·210	0·155	0·115
Time/s	10	20	40	60	100	150	200

Determine the overall order of the reaction and the rate constant. You may assume that 1 mol of N-acetyl cysteine reacts with 1 mol iodoacetamide.

4. In the same reaction as in Question 3 under different conditions the following results were obtained:

Initial concentrations of N-acetyl cysteine and iodoacetamide are 1 mmol dm^{-3} and 2 mmol dm^{-3} respectively.

N-acetyl cysteine concentration/mmol dm^{-3}	0·74	0·58	0·33	0·21	0·12	0·09
Time/s	5	10	25	35	50	60

Assume as before that the stoichiometry is $1:1$ and calculate the rate constant under these conditions.

5. The reaction between two compounds, A and B in solution is first order in B. The following results were obtained in a kinetic experiment at 300 K.

Initial concentration of B $= 1 \cdot 0$ mol dm^{-3}.

Concentration of A/mmol dm^{-3}	1·000	0·692	0·478	0·29	0·158	0·110
Time/s	0	20	40	70	100	120

Determine the order in A and calculate the rate constant for the reaction.

Would the half life of A be different if the initial concentration of B $= 0 \cdot 5$ mol dm^{-3}?

If the activation energy is $83 \cdot 6$ kJ mol^{-1} calculate the *rate* of reaction at 323 K when the concentration of A and B are both $0 \cdot 1$ mol dm^{-3}.

6. Explain what is meant by the steady state approximation. Consider the scheme below for interaction of an enzyme and its substrate.

$$E + S \underset{k_{-1}}{\overset{k_1}{\rightleftharpoons}} ES$$

$$ES \xrightarrow{k_2} E + P$$

Assuming steady state conditions, derive an expression for the rate of formation of product P.

What conditions are necessary for the steady-state approximation to be valid?

What is the *maximum* rate of production of P, if the concentration of enzyme is E_0?

7. The rate constant of a reaction is $15\ (\text{mol dm}^{-3})^{-1}\ \text{min}^{-1}$ at 298 K and $37\ (\text{mol dm}^{-3})^{-1}\ \text{min}^{-1}$ at 308 K. What is the activation energy for the reaction and the rate constant at 283 K?

8. The rate constant for the association of an inhibitor with carbonic anhydrase was studied as a function of temperature, with the results shown below. What is the activation energy for this reaction?

T/K	289·0	293·5	298·1	303·2	308·0	313·5
$10^{-6}k/(\text{mol dm}^{-3})^{-1}\text{s}^{-1})$	1·04	1·34	1·53	1·89	2·29	2·84

9. The variation of rate constant with ionic strength for reactions in water can be derived by considering the Debye–Hückel theory for dilute solutions (p. 162).

$$\log_{10}(k/k_0) = Z_A Z_B \sqrt{(I)}$$

where k = rate constant, k_0 = rate constant at zero ionic strength, Z_A, Z_B are the charges (with signs) on the ionic species involved, and I = ionic strength.

For the reaction

$$Cr(H_2O)_6^{3+} + SCN^- \rightarrow Cr(H_2O)_5SCN^{2+} + H_2O$$

the following data were obtained:

k/k_0	0·87	0·81	0·76	0·71	0·62	0·50
$I(\times 10^3)$	0·4	0·9	1·6	2·5	4·9	10·0

Are these data consistent with this equation?

What is the dependence of the rate of the reverse reaction on ionic strength?

10. The rate of the reaction between an amino acid and trinitrobenzene sulphonate was studied in buffered solutions as a function of the pH of the solution.

pH	6·5	7·0	7·5	8·0	8·5	9·0	9·5	10·0	10·5	11·0	11·5
k_{relative}	0·074	0·25	0·8	2·4	7·5	20	56·6	72	90	97	100

Deduce what you can about the species involved in the reaction. The pK_a of the sulphonic acid group is so low that it remains as the anion throughout the pH range.

11. Nitramide (NH_2NO_2) decomposes in aqueous solution according to the equation

$$NH_2NO_2 \rightarrow N_2O + H_2O.$$

The rate of the reaction (as monitored by the release of N_2O) was studied in acetic acid/acetate buffers of varying composition. The following data were obtained:

pH	Total concentration of acetic acid + acetate ($mmol\ dm^{-3}$)	$10^3\ k\ (min^{-1})$
4·09	18·2	2·12
4·11	20·3	2·46
4·40	20·3	3·82
4·46	26·5	5·26
5·01	20·3	7·26
4·77	28·4	8·00

Given that the pK_a of acetic acid is 4·7 deduce whether the rate constant depends on the concentration of H^+, acetate, or acetic acid.

12. The first-order decomposition of diacetyl ($CH_3COCOCH_3$) has values of $\Delta G^{\circ\ddagger}$ and $\Delta H^{\circ\ddagger}$ of 231·5 kJ mol^{-1} and 264·2 kJ mol^{-1} respectively at 558 K. Calculate the value of $\Delta S^{\circ\ddagger}$ and comment on its magnitude.

10. The kinetics of enzyme catalysed reactions

Introduction

AT first sight the kinetics of enzyme catalysed reactions might appear to be vastly more complex than those of the relatively simple reactions considered in the previous chapter. Enzymes are much more efficient catalysts than any model catalysts yet devised, and in addition the catalytic activity of enzymes is often very sensitive to experimental conditions, such as temperature, pH, and ionic strength. Nevertheless, provided care is taken to obtain reliable and reproducible data (e.g. by the use of buffer solutions at defined temperatures) the kinetics of enzyme catalysed reactions can usually be interpreted in terms of the simple ideas developed so far.

There are at least four reasons why a study of enzyme kinetics is important for the biochemist. Firstly, such a study is observing the enzyme 'doing its job' as a biological catalyst. Secondly, a kinetic study provides us with valuable information about the mechanism of an enzyme catalysed reaction. Thirdly, knowledge of the kinetic parameters (Michaelis constant and maximum velocity) allows us to estimate the importance of a particular enzyme catalysed reaction under the conditions of enzyme and substrate concentration which might prevail in the cell (see, for instance, Problem 2). Fourthly, from measurements of how the rate of an enzyme catalysed reaction is affected by changes in variables such as pH and ionic strength, or by the binding of ligands which may regulate the enzyme activity, we can gain further insight into the likely properties of an enzyme in its environment *in vivo*.

In this chapter we shall mainly be discussing the steady-state behaviour of enzymes under conditions where the enzyme is present in very small concentrations (usually 1 per cent or less) compared to the concentrations of substrates or other ligands.† This makes the algebraic treatment of the various equilibria rather easier (see Problem 2, Chapter 4), since it is then possible to set the *free* substrate concentration equal to the *total* added substrate concentration.

It should be mentioned, however, that many important kinetic studies are now being performed with concentrations of enzyme comparable with those of substrates. Using fast-recording or other special techniques, it then becomes possible to identify intermediate complexes in the overall reaction

---

† The steady-state approximation is valid because the maximum concentration of intermediates (e.g. ES complexes) cannot exceed that of the enzyme, and so the *rate* of change of the concentrations of intermediates is very small compared with the rate of change of the concentrations of substrate or product. See Chapter 9, pp. 151–2.

and to determine the rate constants of some of the individual steps. This type of study will be dealt with in more detail later. In this connection we should note that there are many examples known where enzymes are present *in vivo* in quite high concentrations, and thus it is becoming increasingly clear that for a complete understanding of the rates of enzyme catalysed reactions in the cell we must accumulate kinetic data over as wide a range of enzyme concentrations as possible.

Steady-state kinetics

Initially we shall confine the discussion to one-substrate reactions (i.e. those reactions in which only one substrate is acted on).[†] This will allow us to introduce the basic concepts and definitions which can then be used in the study of enzymes with more than one substrate.

The treatment of enzyme kinetics assumes that a complex between enzyme and substrate is rapidly and reversibly formed.[‡] This complex then breaks down in a slow step to give the product and regenerate enzyme:

$$E + S \underset{k_{-1}}{\overset{k_1}{\rightleftharpoons}} ES$$

$$ES \xrightarrow{k_2} E + P.$$

It has been possible in a number of cases to observe the ES complex directly by using high concentrations of enzyme. Sometimes a reasonably stable intermediate can be isolated (e.g. acyl-chymotrypsin) and this must be incorporated into the overall kinetic scheme.

We can derive the kinetic expression to deal with this pathway in one of two ways.

(1) *The equilibrium approximation*

Here we assume that the $E + S \rightleftharpoons ES$ equilibrium is only slightly disturbed by the breakdown of ES to give the product. This will clearly be a better assumption the lower the value of k_2 relative to k_{-1}.

$$K = \frac{[E][S]}{[ES]}$$

where $[E]$ is the concentration of free enzyme and $[S]$ is the concentration of free substrate. However since $[S_{total}] \gg [E_{total}]$, $[S_{free}] = [S_{total}]$.

[†] It will be assumed that hydrolytic enzymes (i.e. those catalysing hydrolysis reactions) belong to this category, since the second substrate (water) is present in vast excess.

[‡] The part of the enzyme molecule at which the substrate binds is known as the *active site* of the enzyme.

Now the fraction F of enzyme in the form of ES is given by:

$$F = \frac{[ES]}{[E]+[ES]}$$

$$= \frac{[S]}{K+[S]}.$$

Let V_{max} be the maximum rate of product formation, i.e. the rate when all the enzyme is in the form of ES†, then

$$\text{Rate } v = V_{max} \cdot F$$

$$\therefore \quad \boxed{v = V_{max} \cdot \frac{[S]}{K+[S]}}^{‡} . \qquad (10.1)$$

The dependence of v on [S] is illustrated in Fig. 10.1.

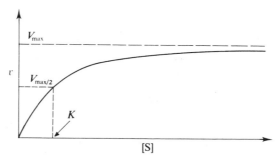

FIG. 10.1. Dependence of the velocity of an enzyme catalysed reaction v on the substrate concentration [S].

When [S] is low compared with K, the reaction is first order in S. At very high [S], the reaction rate tends to V_{max}, and is of zero order in S. In the intermediate range, the reaction is of a fractional order in S.

---

† Under any conditions, of course, $v = k_2[ES]$ (see Chapter 9, p. 152). $V_{max} = k_2[ES]$ when all the enzyme is in the form of ES.

‡ Sometimes this expression is written in the form

$$v = k_{cat} \frac{[E]_0[S]}{K+[S]} \quad \text{where } k_{cat} = V_{max}/[E]_0.$$

$[E]_0$ is the total concentration of enzyme. k_{cat} is thus a first-order rate constant which is equivalent in this mechanism to k_2.

From eqn (10.1) it is clear that when $v = V_{max}/2$, $[S] = K$. K (which thus corresponds to the concentration of substrate when the velocity is half maximal) is known as the Michaelis constant (usually written K_m).† In this case we can equate K with the dissociation constant of the ES complex.

(2) The steady-state approximation

We now abandon the assumption that the $E + S \rightleftharpoons ES$ equilibrium is not disturbed by the breakdown of ES, and use instead the steady-state approximation. After an initial phase (the pre-steady state) it is assumed that the concentration of ES remains constant or very nearly so. In cases where the ES complex can be observed, the experimental data generally resembles that shown in Fig. 10.2 and the rate at which [ES] falls is very low, i.e. the approximation is a valid one for most purposes.

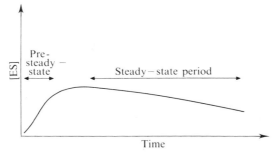

FIG. 10.2. The concentration of the enzyme-substrate complex as a function of time of reaction.

Now in the steady-state, the rate of production of ES (i.e. $k_1[E][S]$) must equal the rate of its breakdown (i.e. $k_{-1}[ES] + k_2[ES]$)

$$\therefore \quad k_1[E][S] = k_{-1}[ES] + k_2[ES]$$

$$\therefore \quad [ES] = \frac{k_1[E][S]}{k_{-1} + k_2}$$

therefore, as before, the fraction (F) of enzyme in the form of the ES complex is given by

$$F = \frac{[ES]}{[E] + [ES]}$$

$$= \frac{[S]}{[(k_{-1} + k_2)/k_1] + [S]}$$

---

† This is the only unambiguous definition of K_m. Any other definition of K_m assumes a particular kinetic scheme.

and hence the rate of the reaction v is given by

$$v = V_{max} \cdot F$$

$$\therefore \quad v = V_{max} \cdot \frac{[S]}{[(k_{-1}+k_2)/k_1]+[S]}. \tag{10.2}$$

Again it should be remembered that $[S] = [S_{total}]$ since the enzyme is assumed to be present in very small amounts.

In the steady-state treatment, the term K in the denominator of eqn (10.1) has been replaced by $(k_{-1}+k_2)/k_1$. Only if $k_2 \ll k_{-1}$, does the term $(k_{-1}+k_2/k_1)$ become equal to k_{-1}/k_1, i.e. to the dissociation constant for the ES complex (K_s). We should therefore find that the Michaelis constant (the concentration of substrate where the velocity is half maximal) is not generally equal to K_s.

From both treatments of the kinetic scheme, we obtain the basic equation†

$$\boxed{v = \frac{V_{max}[S]}{K_m+[S]}} \tag{10.3}$$

where K_m is the Michaelis constant.

Worked example

Using eqn (10.3), calculate the change in [S] required to increase the rate of a reaction from 10 to 90 per cent of the maximum rate. What further change is required to increase the rate to 95 per cent of the maximum rate?

Solution

Put $v = 0 \cdot 1 \ V_{max}$.

Then $\dfrac{[S]}{K_m+[S]} = 0 \cdot 1$,

i.e. $[S] = \dfrac{K_m}{9}$.

Putting $v = 0 \cdot 9 \ V_{max}$ we find $[S] = 9 \ K_m$,

i.e. an 81-fold change in [S] is required.

---

† If we have more than one complex in the pathway, i.e.

$$E+S \rightleftharpoons ES \rightleftharpoons ES' \rightleftharpoons ES'' \rightarrow E+P$$

then applying the steady-state approximation to the various intermediates (ES, ES', etc.) it can be shown that the form of the equation is the same as (10.3). K_m will now be a more complex function of the various rate constants.

Putting $v = 0.95\ V_{max}$

we find $[S] = 19\ K_m$,

i.e. a further 2·11-fold change in [S] is required.

Treatment of kinetic data

From the result of the worked example we see that it is difficult to achieve truly saturating conditions and thus determine V_{max} (and hence K_m) from the plot of v against [S]. To overcome this difficulty we can rearrange eqn (10.3) in a number of ways. Three of the best known rearranged equations are given below.

(1) *The Lineweaver–Burk method*

This uses the rearranged equation

$$\frac{1}{v} = \frac{K_m}{[S]} \cdot \frac{1}{V_{max}} + \frac{1}{V_{max}}\ .$$

Thus a plot of $1/v$ against $1/[S]$ gives a straight line whose intercepts on the x and y axes are $(-1/K_m)$ and $1/V_{max}$ respectively, and whose slope is (K_m/V_{max}) (Fig. 10.3).

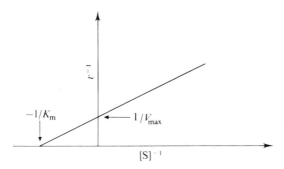

FIG. 10.3. The Lineweaver–Burk method of plotting enzyme kinetic data.

It will be noticed that this plot is rather similar to that used in the analysis of binding data (the Hughes–Klotz plot, Fig. 4.3).

(2) *The Eadie–Hofstee method*

The basic equation is rearranged to give

$$\frac{v}{[S]} = \frac{V_{max}}{K_m} - \frac{v}{K_m} \quad .$$

A plot of $v/[S]$ against v gives a straight line with a slope of $-1/K_m$ and an intercept on the x-axis of V_{max} (Fig. 10.4). (This type of plot is analogous to the Scatchard plot for binding data (Fig. 4.4).)

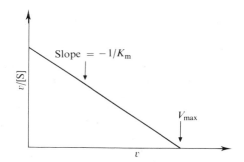

FIG. 10.4. Eadie–Hofstee method of plotting enzyme kinetic data.

(3) *The Hanes method*

Eqn (10.3) is rearranged to give

$$\frac{[S]}{v} = \frac{[S]}{V_{max}} + \frac{K_m}{V_{max}} \quad .$$

A plot of $[S]/v$ against $[S]$ is linear with a slope of $1/V_{max}$ and an intercept on the x-axis of $-K_m$. (Fig. 10.5.)

Which plot to use?

This question (i.e. which plot should be used to derive kinetic parameters from a set of data?) has generated a good deal of discussion among biochemists. Undoubtedly the Lineweaver–Burk plot is the most commonly used and it has the merit that the variables (v and $[S]$) are plotted on separate axes. On the other hand an analysis of the errors involved in the

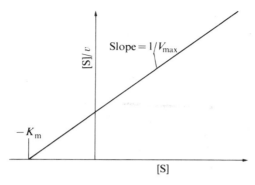

FIG. 10.5. Hanes method of plotting enzyme kinetic data.

determination of the experimental data and the subsequent extraction of the parameters K_m and V_{max} has shown that the distribution of errors is highly non-uniform over the range of values of v and [S] in the Lineweaver–Burk plot. The error distribution is more uniform in the other two plots mentioned and so the use of these plots has been advocated by a number of workers, since statistical analysis of the data is more reliable. (In particular, the use of the Hanes plot has been recommended, since in this plot the error present in the x-axis is likely to be very small.) What should be remembered, however, is that any serious kineticist uses a computer method of fitting the data to the original rectangular hyperbola (eqn 10.3) to obtain the 'best fit' values of K_m and V_{max}, and that the linear plots are then merely representations of the data in an 'easy to comprehend' form. A fuller discussion of the merits of various plots is given in the books by Cornish-Bowden mentioned in the reading list.

A note on units

The units of K_m are those of concentration (i.e. $mol\ dm^{-3}$)† as K_m is the *concentration* of substrate at which half maximal velocity is observed.

V_{max} can be expressed in a variety of units depending on what information is available. If we are working with an impure preparation of enzyme or if a purified enzyme has not been characterized as far as its molecular weight is concerned, the velocity (v) will generally be given in units of moles substrate consumed (or product formed) per weight of protein (or enzyme) per unit time (e.g. $1\ mmol\ (g\ protein)^{-1}\ s^{-1}$). The SI unit of enzyme activity is the *katal* which is defined as:

---

† Strictly speaking K_m as a type of equilibrium constant should be dimensionless (Chapter 3). However, biochemists invariably quote units, e.g. $K_m = 50\ \mu mol\ dm^{-3}$.

One katal is that amount of enzyme which, under the specified conditions, catalyses the production of 1 mole product (or the consumption of 1 mole substrate) per second.

The *specific activity* of an enzyme would then be expressed in units of katal $(kg\ enzyme)^{-1}$ (or katal $(kg\ protein)^{-1}$ for an impure enzyme preparation).

If, however, we know the concentration of enzyme active sites (in molar terms), we can evaluate k_{cat} for the reaction ($k_{cat} = V_{max}/[E]_0$; see footnote on p. 170) which is also known as the *turnover number* of the enzyme (moles substrate consumed or product formed per mole enzyme per unit time).

Some practical aspects

In order to obtain reliable results in enzyme kinetic work, it is necessary to take a number of experimental precautions of which we might mention the following:

(i) The substrates, buffers, etc. should be of as high a purity as possible, since contaminating substances may affect the activity of the enzyme. (Commercial preparations of NAD^+ sometimes contain inhibitors of dehydrogenases, for example.)

(ii) It must be ascertained that the enzyme preparation does not contain any substances (or other enzymes) which interfere with the assays of activity. Also the possibility of non-enzyme catalysed reactions should be tested by appropriate control experiments (e.g. by addition of heat-inactivated enzyme).

(iii) The enzyme should be stable (i.e. not lose any significant activity) under the conditions chosen, for at least the length of time required to perform the assays.

(iv) It should be checked that the measured rate is proportional to the amount of enzyme added and that the rate of product formation (or substrate consumption) is linear over the period of interest.

(v) The rate can be measured more accurately (and conveniently) by making a continuous record of product formation or substrate consumption rather than by performing analysis of the reaction mixture at fixed time intervals. A continuous recording is easily made, for instance, using a spectrophotometer if the reaction involves a change in light absorption (Chapter 11).

If, however, the reaction of interest does not involve a convenient change in light absorption or some other readily measurable property this difficulty can sometimes be overcome by using a *coupled-assay procedure*, e.g. the reaction catalysed by pyruvate kinase:

$$\text{phosphoenolpyruvate} + \text{ADP} \xrightarrow{\text{Mg}^{2+}} \text{pyruvate} + \text{ATP}$$

is not easily monitored spectrophotometrically, but if NADH and lactate dehydrogenase are added to the reaction mixture, the pyruvate formed is reduced:

$$\text{pyruvate} + \text{NADH} \rightarrow \text{lactate} + \text{NAD}^+$$

and this latter reaction is easily followed since NADH absorbs radiation of wavelength 340 nm but NAD^+ does not.

It is essential to add the coupling enzyme(s) and substrate(s) in sufficient quantities to ensure that the overall measured rate is the rate of the first (pyruvate kinase) reaction, i.e. that the pyruvate as soon as it is formed is converted to lactate with concomitant oxidation of NADH.

Worked examples

(1) Under certain conditions the maximum specific activity of a purified preparation of triosephosphate isomerase was quoted as 10 000 μmoles glyceraldehyde-3-phosphate transformed per mg protein per minute. Express this in terms of katal kg^{-1} and evaluate the turnover number of the enzyme given that its active site molecular weight is 26 500.

Solution

10 000 $\mu\text{mol mg}^{-1}\,\text{min}^{-1}$ is equivalent to $\dfrac{10\,000}{60}\,\mu\text{mol mg}^{-1}\,\text{s}^{-1}$,

i.e. to $\dfrac{10\,000}{60} \times 10^6\,\mu\text{mol kg}^{-1}\,\text{s}^{-1}$,

i.e. to $\dfrac{10\,000}{60}\,\text{mol kg}^{-1}\,\text{s}^{-1}$,

i.e. $\underline{166 \cdot 7\ \text{katal kg}^{-1}}$.

Now 1 mole enzyme (active sites) $= 26 \cdot 5$ kg

therefore turnover number $= 166 \cdot 7 \times 26 \cdot 5\ \text{s}^{-1}$

$$= 4420\ \text{s}^{-1}.$$

(2) The activity of the enzyme urease, which catalyses the reaction

$$\text{CO(NH}_2)_2 + \text{H}_2\text{O} \rightleftharpoons \text{CO}_2 + 2\text{NH}_3$$

was studied as a function of urea concentration with the following results:

Urea concentration (mmol dm^{-3})	30	60	100	150	250	400
Velocity (mmol urea consumed (mg enzyme)$^{-1}$ min^{-1})	3·37	5·53	7·42	8·94	10·70	12·04

What are the values of K_m and V_{max} for this reaction?

Solution

The data are rearranged into an appropriate form for a linear plot (i.e. $1/v$ vs. $1/[S]$, $v/[S]$ vs. v, $[S]/v$ vs. $[S]$).

These three plots are shown in Fig. 10.6. From all three plots we deduce that the K_m is <u>105 mmol dm^{-3}</u> and the V_{max} is <u>15·2 mmol urea consumed mg^{-1} min^{-1}</u> (this is equivalent to <u>253·3 katal kg^{-1}</u>. Of course we would not normally do all three plots; this was done here merely for illustrative purposes. It might be noted that the distributions of points are markedly different in the three plots and this fact should be borne in mind when suitable values for substrate concentration are being chosen for such an experiment.

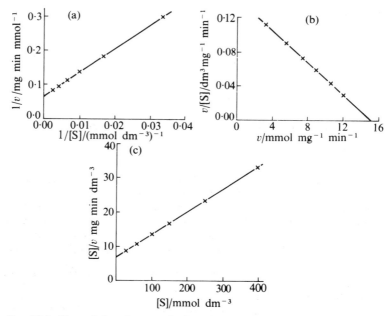

FIG. 10.6. Plots of data from worked example according to the methods of (*a*) Lineweaver–Burk, (*b*) Eadie–Hofstee, (*c*) Hanes.

Complications to the basic equation

We should note that an equation such as (10.3) is an *initial-rate equation* (i.e. the concentration of S is assumed to remain at its starting value). As the concentration of S falls during the course of a reaction, the rate will, of course, decline too. Also we do not have to consider the reverse reaction or any possible inhibition by product.† However, deviations from these initial rate equations can occur for a number of reasons of which we might mention two of the most important.

(a) *Substrate inhibition*

This occurs with some enzymes at high substrate concentrations and makes the linear plots curved‡ (Fig. 10.7 shows this for a Lineweaver–Burk plot).

An example of this type of effect is seen in the reaction catalysed by insect acetylcholinesterase   (acetylcholine + water → acetate + choline)   where marked inhibition occurs at acetylcholine concentrations above about 1 mmol dm^{-3}.

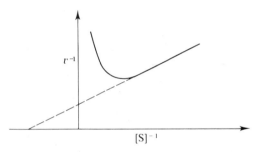

FIG. 10.7. Lineweaver–Burk plot for the case of substrate inhibition.

(b) *Multiple binding*

Complex kinetics are often observed when an enzyme is composed of multiple subunits (and therefore possesses a number of active sites). We have already discussed this type of system in connection with ligand binding (Chapter 4) and, although the equations involved are complex,† we should note that the treatment of enzyme kinetics has many analogies with the treatment of ligand binding data. The effects of interaction between

---

† These aspects are dealt with more fully in, for example, the book by Ferdinand in the reading list.

‡ Such behaviour is usually interpreted in terms of a second site for substrate binding (which becomes significantly occupied only at high substrate concentrations). Occupation of this second site inhibits the reaction occurring at the first (catalytic) site.

subunits (which can be manifested as either positive or negative co-operativity) are often of considerable importance in the regulation of enzyme activity, since they render the enzyme activity more sensitive or less sensitive to changes in [S] than is the case with hyperbolic binding† (eqn (10.3).

Enzyme inhibition

A very important aspect of the study of enzyme catalysed reactions is a study of the effects of inhibitors since this can give information on the active site of an enzyme, its mechanism of action, and possible regulation of physiological significance. We will discuss this in terms of the scheme shown in Fig. 10.8. (Enzyme inhibition can also be discussed in terms of other, more complex schemes but we shall confine ourselves to the simplest cases only.)

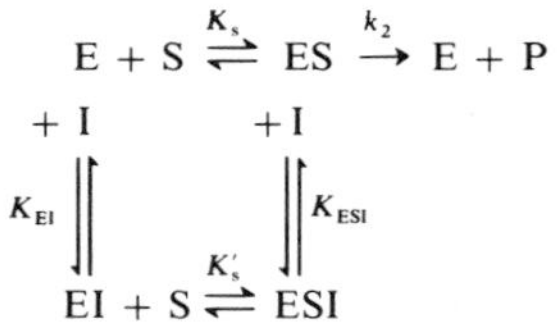

$$
\begin{array}{ccc}
\mathrm{E + S} & \overset{K_s}{\rightleftharpoons} \mathrm{ES} & \overset{k_2}{\to} \mathrm{E + P} \\
+\,\mathrm{I} & +\,\mathrm{I} & \\
K_{\mathrm{EI}} \Big\updownarrow & \Big\updownarrow K_{\mathrm{ESI}} & \\
& \overset{K_s'}{} & \\
\mathrm{EI + S} & \overset{}{\rightleftharpoons} \mathrm{ESI} &
\end{array}
$$

FIG. 10.8. The general scheme for enzyme inhibition. K_s, K_{EI}, etc., represent dissociation constants. The breakdown of ES yields product (ESI is assumed to be inactive).

We shall assume that all the enzyme-containing complexes are in equilibrium with each other (i.e. that the k_2 step does not significantly disturb these equilibria). The general equation for this situation‡ is (see Appendix 4)

$$
\frac{1}{v} = \frac{1}{V_{\max}}\left(1 + \frac{[\mathrm{I}]}{K_{\mathrm{ESI}}}\right) + \frac{K_s}{V_{\max}}\left(1 + \frac{[\mathrm{I}]}{K_{\mathrm{EI}}}\right)\frac{1}{[\mathrm{S}]} \tag{10.4}
$$

where [S] and [I] refer to substrate and inhibitor respectively. (If we use the steady-state approximation to treat this scheme, the equations derived are more complex, and of little use experimentally.)

---

† Positive or negative co-operativity can be recognized from the plots of v vs. [S] or from the rearranged plots described in this chapter. (The shapes are similar to those mentioned in Chapter 4 in connection with binding data.) An example of positive co-operativity is found in the case of aspartate transcarbamylase which shows a sigmoidal v vs. [S] curve for aspartate. An example of negative co-operativity is seen with the substrate NAD^+ for beef-liver glutamate dehydrogenase.

‡ Since we have made the 'equilibrium' assumption, we can equate K_s with K_m (see p. 171).

The general expression (eqn 10.4) can be simplified by making certain assumptions. These will be discussed as three limiting cases, but it should be noted that several other cases are possible.

(1) Competitive inhibition

If, in eqn (10.4), $K_{ESI} = \infty$ (i.e. the ES complex cannot combine with I, nor the EI complex with S), the general equation simplifies to:

$$\frac{1}{v} = \frac{1}{V_{max}} + \frac{K_s}{V_{max}}\left(1 + \frac{[I]}{K_{EI}}\right)\frac{1}{[S]} \tag{10.5}$$

This is the situation known as *competitive inhibition*. The effect on the Lineweaver–Burk and Eadie–Hofstee plots is shown in Fig. 10.9 (the effect on the Hanes plot is left as an exércise in Problem 1).

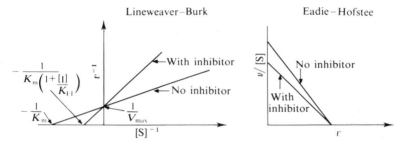

FIG 10.9. The effect of a competitive inhibitor on the enzyme kinetic plots.

Effectively, the inhibitor 'pulls' some of the enzyme over into the form of the EI complex. When the concentration of S is increased sufficiently, this effect can be overcome.† Thus V_{max} remains the same, but K_m is increased by the factor $(1 + [I]/K_{EI})$.

Many examples of competitive inhibition are known, e.g. carbamyl-choline

$$(CH_3)_3\overset{\oplus}{N}-CH_2CH_2-O-\overset{\overset{\displaystyle O}{\|}}{C}-NH_2$$

and several other compounds containing a quaternary N atom act as competitive inhibitors with respect to acetylcholine

$$(CH_3)_3\overset{\oplus}{N}-CH_2CH_2-O-\overset{\overset{\displaystyle O}{\|}}{C}-CH_3$$

---

† This is the basis of the term 'competitive', i.e. at high [S], the effect of I on the reaction velocity can be overcome. With a 'non-competitive' inhibitor, as we shall see, the effect of I on the reaction velocity cannot be overcome by increasing [S].

in the reaction catalysed by acetylcholinesterase from bovine eryth-rocytes.† It might be thought that, if competitive inhibition were observed, it would prove that the inhibitor and substrate bind to the same site. While in many cases the structural analogy between the competitive inhibitor and the substrate may make this very likely, the conclusion is not necessarily justified. However, it can be said that if (from other data) S and I are known to bind to the same site on the enzyme, then competitive inhibition will be observed.

(2) *Non-competitive inhibition.*

If, in eqn (10.4) $K_{ESI} = K_{EI}$ (i.e. the binding of S to the enzyme does not affect the binding of I) then

$$\frac{1}{v} = \frac{1}{V_{max}}\left(1 + \frac{[I]}{K_{EI}}\right) + \frac{K_s}{V_{max}}\left(1 + \frac{[I]}{K_{EI}}\right) \cdot \frac{1}{[S]} \tag{10.6}$$

This is known as *non-competitive inhibition.* K_m remains unaffected, whereas V_{max} is decreased by a factor $1/\{1 + ([I]/K_{EI})\}$ as is shown in the plot (Fig. 10.10)

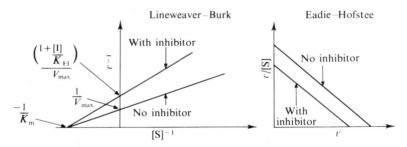

FIG. 10.10. The effect of a non-competitive inhibitor on the enzyme kinetic plots.

Examples of non-competitive inhibition are rather less common than examples of competitive inhibition in the case of one substrate reactions (although there are many examples of non-competitive inhibition in the case of multi-substrate reactions). In one substrate reactions an example is the inhibition of the chymotrypsin-catalyzed hydrolysis of N-acetyl-L-tyrosine ethylester by indole (which behaves as a non-competitive inhibitor towards the substrate).

---

† Many of these compounds have important pharmacological applications since acetylcholine is involved in nervous transmission.

(3) *Uncompetitive inhibition*

If, in eqn (10.4) $K_{EI} = \infty$ (i.e. E cannot combine with I) then

$$\frac{1}{v} = \frac{1}{V_{max}}\left(1 + \frac{[I]}{K_{ESI}}\right) + \frac{K_s}{V_{max}}\frac{1}{[S]} \qquad (10.7)$$

This is known as simple *uncompetitive inhibition*. Both K_m and V_{max} are affected, as is shown in the plots (Fig. 10.11). Examples of simple uncompetitive inhibition are extremely rare in one substrate reactions (an example is the inhibition of rat intestinal alkaline phosphatase by L-phenylalanine) but are more often found in multisubstrate reactions (e.g. S-adenosylmethionine behaves as an uncompetitive inhibitor towards ATP in the reaction catalysed by yeast S-adenosylmethionine synthase).

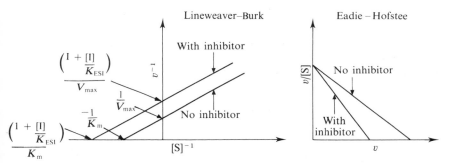

FIG. 10.11. The effect of a simple uncompetitive inhibitor on the enzyme kinetic plots.

It should perhaps be noted that the net result of a non-competitive inhibitor (i.e. a decrease in V_{max}, but the same K_m) is *equivalent* to converting some of the enzyme present to an inactive form. Thus, for instance, diisopropylfluorophosphate behaves as a non-competitive inhibitor towards the substrates of chymotrypsin since it inactivates the enzyme by reacting with an essential serine group. We shall restrict the terms 'competitive', 'non-competitive', and 'uncompetitive' to those inhibitors which bind *reversibly* to any enzyme. It is also more useful, in our view, to use these terms to describe the effects that an inhibitor shows on the kinetic plots (since these can be readily determined), rather than to use them to describe the probable relationship between inhibitor and substrate binding sites on the enzyme (since this can be very difficult to establish).

It is worth re-emphasizing that in many cases (especially in the cases of multisubstrate reactions or multiple binding sites) the effects of an inhibitor do not conform exactly to any of the limiting situations discussed here. Such 'mixed' inhibition patterns clearly require different assumptions and equations for their description.

Worked example

The effect of choline on the reaction catalysed by insect acetyl-cholinesterase was studied with the following results:

[choline] (mmol dm^{-3})	[Acetylcholine] (mmol dm^{-3})				
	0·1	0·15	0·25	0·40	0·70
	(relative velocity (arbitary units))				
0	28·5	37·5	50·0	61·5	73·7
20	11·9	15·6	20·9	25·7	30·7
40	7·5	9·9	13·2	16·2	19·4

What type of inhibition is being observed in this case?

Solution

Plotting the data in a suitable form (Fig. 10.12 shows a Lineweaver–Burk plot but any of the three plots is suitable for this purpose) we note that K_m is unchanged (0·25 mmol dm^{-3}) in the presence of choline whereas V_{max} is

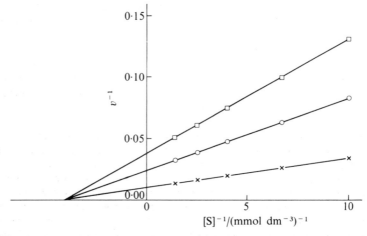

FIG. 10.12. Lineweaver–Burk plot of data from worked example. Data in the presence of 0, 20, and 40 mmol dm^{-3} choline are shown by ($\times$), ($\bigcirc$), and ($\square$) respectively.

lowered (the values in the presence of 0, 20 and 40 mmol dm^{-3} choline are 100·0, 41·7, and 26·3 respectively. From eqn (10.6) we might note that $V_{max}/(V_{max})_I = 1 + [I]/K_{EI}$, i.e. a plot of $V_{max}/(V_{max})_I$ against [I] should be linear with an intercept on the y-axis of 1. This is found to be the case here and the value of K_{EI} can be evaluated as 14·3 mmol dm^{-3}.

The non-competitive inhibition can be interpreted in terms of a second site for choline (i.e. distinct from the active site) and the note on substrate inhibition with this enzyme should be consulted (p. 179).

Two substrate kinetics

One substrate kinetics, which we have been discussing up to now, is only of limited applicability as can be seen by considering the types of reactions which enzymes catalyse. Any given enzyme can be placed into one of six categories according to the reaction type.

(1) *Oxidoreductases.*	Catalyse redox reactions in which one substrate is reduced at the expense of a second which is oxidized.
(2) *Transferases.*	Catalyse reactions in which a group is transferred from one substrate to another.
(3) *Hydrolases.*	Catalyse reactions in which a substrate is hydrolysed.
(4) *Lyases.*	Catalyse reactions in which a group is eliminated from a substrate to form a double bond.
(5) *Isomerases.*	Catalyse isomerization reactions.
(6) *Ligases.*	Catalyse the joining together of two molecules at the expense of ATP, or some other 'energy source'.

Enzymes belonging to categories (4) and (5) can be considered as one substrate enzymes, and hydrolases (category (3)) can also be included under this heading, since the second substrate (water) is present in vast excess. However, enzymes belonging to the other three categories clearly catalyse reactions of more than one substrate and it is these reactions that we shall turn to now. We shall indicate the various types of mechanism which can occur, discuss the resulting equations and outline how distinctions between different mechanisms can be achieved. More detailed treatments of this topic can be found in the books by Cornish-Bowden, Engel, and Ferdinand in the reading list.

Types of mechanisms

A basic division in two substrate reactions can be made into the following categories:

(a) *Those involving a ternary complex*

These reactions proceed via EAB and EPQ complexes (A and B are the substrates, and P and Q are the products)

$$E + A + B \rightarrow EAB$$
$$EAB \rightarrow EPQ$$
$$EPQ \rightarrow E + P + Q$$

This category can be subdivided into

(i) Those reactions in which the ternary complex (EAB) is formed in an *ordered* manner, i.e.

$$E + A \rightarrow EA$$
$$EA + B \rightarrow EAB \quad (\text{but } E + B \nrightarrow EB).$$

(ii) Those reactions in which the ternary complex is formed in a *random* manner, i.e.

$$E + A \rightarrow EA \qquad E + B \rightarrow EB$$
$$or$$
$$EA + B \rightarrow EAB \qquad EB + A \rightarrow EAB$$

(b) *Those not involving a ternary complex*

This category can be subdivided into

(i) Those reactions in which the first product is formed before the second substrate is bound. These cases involve a modification of the enzyme and are known as *enzyme substitution* or *ping-pong* mechanisms.

$$E + A \rightarrow E' + P$$
$$E' + B \rightarrow E + Q$$

(ii) Those reactions in which a ternary complex is presumably formed, but its breakdown to yield the first product is very fast (so that the ternary complex is kinetically insignificant). This is the 'Theorell–Chance' mechanism and holds for the oxidation of ethanol catalysed by horse-liver alcohol dehydrogenase.

Kinetic equations for two substrate reactions

We shall not go into the details of the derivation of the various equations for these possible mechanisms (these are covered in the various books already mentioned). However, these derivations do not involve any new fundamental principles. We apply the steady-state approximation† to evaluate the concentrations of the various enzyme containing complexes in the reaction pathway, and then the velocity of the overall reaction is set equal to the concentration of the complex preceding enzyme-regeneration multiplied by the rate constant for the step which regenerates enzyme.

The equations which result from this treatment are of the following form.

For the ternary complex mechanisms‡

$$v = \frac{V_{max}\,[A][B]}{K'_A K_B + K_B[A] + K_A[B] + [A][B]} \tag{10.8}$$

where K'_A, K_A, and K_B are constants, the meaning of which we shall discuss shortly.

For the ping-pong mechanism

$$v = \frac{V_{max}[A][B]}{K_B[A] + K_A[B] + [A][B]}. \tag{10.9}$$

Significance of the parameters in the kinetic equations

V_{max} in eqns (10.8) and (10.9) represents the maximum velocity at saturating levels of A and B.

In an *operational* sense, the parameters K_A and K_B in eqns (10.8) and (10.9) represent the Michaelis constants for each substrate (A and B respectively) in the presence of saturating concentrations of the other substrate. This can be readily shown, e.g. in eqn (10.8) by dividing numerator and denominator by [B] and then setting $[B] \to \infty$. The

---

† If we apply the steady-state approximation in the case of the random order ternary complex mechanism (*a*)(ii), the equation which results is rather complex and contains terms in the square of the concentrations of substrates. It is possible however to apply the *equilibrium approximation* (as with one substrate kinetics) and assume that E, EA, EB, and EAB are all in equilibrium with each other. The resulting equation (10.8) describes a number of reactions of this type and to that extent the equilibrium assumption is justified.

‡ An equation of this form is also derived for the Theorell–Chance mechanism (*b*)(ii).

equation reduces to:

$$v = \frac{V_{max}[A]}{K_A + [A]},$$

i.e. K_A is the Michaelis constant for A at saturating levels of B.

A similar procedure, dividing eqn (10.8) through by [A] shows that K_B is the Michaelis constant for B at saturating levels of A. K'_A in eqn (10.8) does not have a similarly simple operational meaning.

In terms of the *mechanisms* of the reactions, K'_A, K_A and K_B represent combinations of rate constants of individual steps in the reactions. Their precise meaning varies according to the type of mechanism under discussion. In the case of the random-order ternary complex mechanisms (a)(ii), however, K'_A, K_A, and K_B have relatively simple meanings in terms of equilibrium constants:†

$$\left(\text{note } K'_B = \frac{K'_A K_B}{K_A}\right).$$

Derivation of the kinetic parameters from experimental data

The four parameters in eqn (10.8) can be derived in the following fashion. Values of the velocity, v, are measured at various values of [A] with the concentration of B constant. The procedure is then repeated at other fixed values of [B].

By taking the inverse of eqn (10.8), i.e.

$$\frac{1}{v} = \left(1 + \frac{K_A}{[A]} + \frac{K_B}{[B]} + \frac{K'_A K_B}{[A][B]}\right)\frac{1}{V_{max}} \qquad (10.10)$$

we see that *primary plots* of $1/v$ vs. $1/[A]$ (at constant values of [B]) will be linear with

$$\text{a } slope \text{ of } \quad \frac{1}{V_{max}}\left[K_A + \frac{K'_A K_B}{[B]}\right] \qquad (10.11)$$

$$\text{and an } intercept \text{ on the } y\text{-}axis \text{ of } \quad \frac{1}{V_{max}}\left[1 + \frac{K_B}{[B]}\right] \qquad (10.12)$$

---

† This is true, of course, if we have made the assumption that the complexes E, EA, EB, and EAB are in equilibrium with each other.

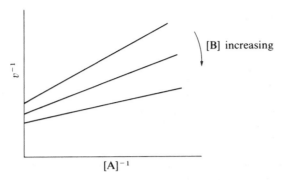

FIG. 10.13. Primary plots of $1/v$ vs. $1/[A]$ at various fixed values of $[B]$, according to eqn (10.10).

The primary plots are shown in Fig. 10.13.†

Thus as $[B]$ *increases*, both the *slope* and the *intercept* will *decrease*.

Secondary plots of the slopes and intercepts of the primary plot vs. $1/[B]$ can then be made (Fig. 10.14).

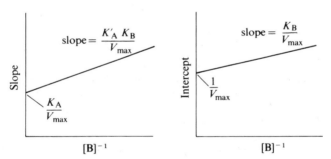

FIG. 10.14. Secondary plots of slopes and intercepts of primary plot (Fig. 10.13) vs. $1/[B]$.

Reference to eqn (10.11) shows that a plot of the slopes vs. $1/[B]$ is linear with a *slope* of $K'_A K_B / V_{max}$ and an *intercept on the y-axis* of K_A / V_{max}. From eqn (10.12) it is seen that a plot of the intercepts vs. $1/[B]$ is linear with a

---

† The lines in the primary plot (Fig. 10.13) intersect at a point which can be above, on, or below the x-axis, depending on the values of K'_A, K_A, and K_B. At any value of $[B]$, an 'apparent' K_m for $[A]$ can be derived from the intercept on the x-axis: obviously this 'apparent' K_m will vary with $[B]$ unless the point of intersection is on the x-axis. The reader can satisfy himself that this case arises when $K'_A = K_A$.

slope of K_B/V_{max} and an *intercept on the y-axis* of $1/V_{max}$. Thus all four parameters in eqn (10.8) can be derived from the slopes and intercepts of these secondary plots.

If a similar procedure is adopted for the ping-pong mechanism (for which eqn (10.9) holds) the resulting primary plot is seen to consist of a series of parallel lines (Fig. 10.15).

This is readily seen by considering the inverse of eqn (10.9), i.e.

$$\frac{1}{v} = \left(1 + \frac{K_A}{[A]} + \frac{K_B}{[B]}\right)\frac{1}{V_{max}}$$

Thus a plot $1/v$ vs. $1/[A]$ will have a slope of K_A/V_{max}. This slope is independent of $[B]$, resulting in a set of parallel lines.†

Worked example

The following data were derived from a study of the reaction catalysed by yeast alcohol dehydrogenase

$$\text{ethanol} + \text{NAD}^+ \rightleftharpoons \text{acetaldehyde} + \text{NADH}$$

[Ethanol] (mmol dm^{-3})	[NAD$^+$] (mmol dm^{-3})			
	0·05	0·1	0·25	1·0
	Velocity (katal kg^{-1})			
10	0·30	0·51	0·89	1·43
20	0·44	0·75	1·32	2·11
40	0·57	0·99	1·72	2·76
200	0·76	1·31	2·29	3·67

Determine the kinetic parameters for this reaction.

Solution

A primary plot $1/v$ vs. $1/[\text{NAD}^+]$ is made for the various values of [ethanol]. This is shown in Fig. 10.16(a).

From the primary plot we can tabulate the values of slopes and intercepts:

[Ethanol] (mmol dm^{-3})	10	20	40	200
Slope	0·139	0·095	0·073	0·055
Intercept	0·55	0·38	0·28	0·22

---

† It is left as an exercise to the reader to show that a secondary plot of the y-axis intercepts of the primary plot vs. $1/[B]$ will have a slope of K_B/V_{max} and a y-axis intercept of $1/V_{max}$. Thus the three parameters in eqn (10.9) can be derived from this secondary plot and the *slope* of the primary plot.

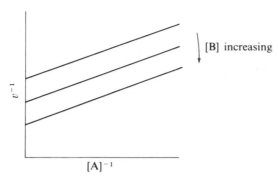

FIG. 10.15. Primary plot of $1/v$ vs. $1/[A]$ at various fixed values of [B], according to eqn (10.9) (the 'ping-pong' mechanism).

The secondary plots (slopes and intercepts vs. $[\text{ethanol}]^{-1}$) are shown in Fig. 10.16(b) and (c).

From the y-axis intercept of Fig. 10.16(c) (0·2) we deduce that $V_{\max} = 1/0·2 = 5$.

From the slope of Fig. 10.16(c) (3·5) we deduce that $K_{\text{ethanol}} = 5 \times 3·5 = 17·5$ (mmol dm^{-3}).

From the y-axis intercept of Fig. 10.16(b) (0·05) we deduce that $K_{\text{NAD}^+} = 5 \times 0·05 = 0·25$ (mmol dm^{-3}).

From the slope of Fig. 10.16(b) (0·89) we deduce that $K'_{\text{NAD}^+} = 5 \times 0·89/17·5 = 0·25$ (mmol dm^{-3}).

So the parameters† are

$V_{\max} = 5$ katal kg^{-1},	$K_{\text{NAD}^+} = 0·25$ mmol dm^{-3}
$K'_{\text{NAD}^+} = 0·25$ mmol dm^{-3},	$K_{\text{ethanol}} = 17·5$ mmol dm^{-3}.

Distinction between the various mechanisms

So far we have seen how to distinguish a 'ping-pong' mechanism from one involving a ternary complex (although it must be emphasized that care should be taken to ensure that lines in a primary plot such as Fig. 10.15 are truly parallel). To confirm that a particular reaction follows a 'ping-pong' mechanism we should look for (i) the occurence of half-reactions, i.e. the

---

† It will be noted that $K_{\text{NAD}^+} = K'_{\text{NAD}^+}$ in this case. This could have been deduced from the primary plot since the lines appear to intersect on the x-axis. However, it is more instructive to go through the secondary plot procedure. Other evidence suggests that this enzyme obeys a random order ternary complex mechanism; thus we can assign these Ks to equilibrium constants. In this case it appears that the binding of NAD$^+$ is not affected by the binding of ethanol and vice versa.

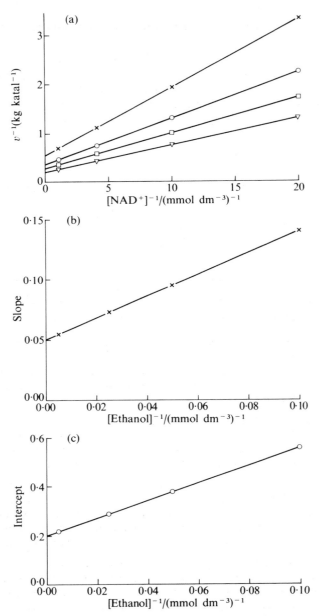

FIG. 10.16. Plot of data from worked example. (a) Primary plot for data at ethanol concentrations of 10 (×), 20(○), 40 (□), and 200 (▽) mmol dm^{-3} respectively. (b), (c) Plot of slopes (×) and intercepts (○) of primary plots against reciprocal of ethanol concentration.

formation of the first product in the absence of the second substrate, and (ii) the formation of a modified enzyme (E′ in the scheme on p. 186), on incubation of the enzyme with the first substrate. Classical examples of 'ping-pong' mechanisms are found among the transaminases in which the pyridoxal prosthetic group on the enzyme becomes modified during the course of the reaction:

$$
\overset{\overset{\displaystyle \overset{\oplus}{N}H_3}{|}}{R-CH-CO_2^{\ominus}} + E-CHO \;\rightleftharpoons\; R-\overset{\overset{\displaystyle O}{\|}}{C}-CO_2^{\ominus} + E-CH_2-\overset{\oplus}{N}H_3^{\oplus}
$$

$$
R'-\overset{\overset{\displaystyle O}{\|}}{C}-CO_2^{\ominus} + E'-CH_2-\overset{\oplus}{N}H_3 \;\rightleftharpoons\; R'-\overset{\overset{\displaystyle \overset{\oplus}{N}H_3}{|}}{CH}-CO_2^{\ominus} + E-CHO
$$

Overall

$$
R-\overset{\overset{\displaystyle \overset{\oplus}{N}H_3}{|}}{CH}-CO_2^{\ominus} + R'-\overset{\overset{\displaystyle O}{\|}}{C}-CO_2^{\ominus} \;\rightleftharpoons\; R-\overset{\overset{\displaystyle O}{\|}}{C}-CO_2^{\ominus} + R'-\overset{\overset{\displaystyle \overset{\oplus}{N}H_3}{|}}{CH}-CO_2^{\ominus}
$$

(E—CHO represents enzyme with pyridoxal phosphate prosthetic group.)

On the other hand, from an analysis of the steady-state kinetics we are unable to distinguish an ordered ternary complex mechanism from a random one, since both give rise to an equation of the form of eqn (10.8).† We can make the distinction however on the basis of other data which might include the following:

(a) *product inhibition patterns.* The type of inhibition shown by products P and Q towards substrates A and B (i.e. is P competitive, non-competitive or uncompetitive with respect to A and/or B?) can be used to indicate a likely mechanism or exclude another.

(b) *substrate binding.* In a random order mechanism, substrates A and B should both be able to bind to the enzyme, whereas in the ordered mechanism the second substrate (say, B) cannot bind in the absence of the first substrate A. Determination of whether the enzyme can bind B or not can help to indicate which mechanism is correct.

(c) *isotope exchange at equilibrium.* This type of experiment involves measuring the influence of the concentration of substrates (and products) on the rate of exchange of an isotope between substrates

---

† Although the Theorell–Chance mechanism (b) (ii) also gives an equation of the form of eqn (10.8), it can be distinguished from the others by comparison of the magnitudes of various parameters in the rate equation with the rates of the forward and reverse reactions. A fuller account of this is given in the book by Engel.

and products when the reaction is at equilibrium. Ordered and random mechanisms can be distinguished on this basis.

More complete discussions on these aspects are given in the books by Engel and Ferdinand. However, we might mention that as a result of studies such as these, it has been shown that for instance muscle creatine kinase and yeast alcohol dehydrogenase proceed via random ternary complex mechanisms and malate dehydrogenase and lactate dehydrogenase proceed via ordered ternary complex mechanisms.

Pre-steady-state kinetics

Up to now we have concentrated on the kinetic behaviour of enzymes when a steady-state has been achieved in reactions where $[E] \ll [S]$. However, a great deal of useful information can be obtained from studies of the pre-steady-state period. Usually in these experiments we make the concentrations of enzyme and substrate more nearly equal and special rapid mixing and detection apparatus is required to initiate and monitor the progress of the reaction. Processes with a half-time of the order of a few milliseconds† can be followed using suitably designed apparatus (the limiting factor is usually the rate at which two solutions can be properly mixed). A full account of these types of study is given in the books by Fersht and Gutfreund.

Among the types of information available from these studies are the following:

(i) The participation of complexes and intermediates in the reaction pathway can be deduced and the rate constants of the individual steps in the pathway evaluated. Thus the participation of an enzyme substrate complex $(E-H_2O_2)$ was demonstrated in the peroxidase catalysed oxidation of leucomalachite green by hydrogen peroxide. In the case of the reaction catalysed by pig-heart lactate dehydrogenase (where steady-state kinetics had indicated an ordered ternary complex mechanism with NAD^+ binding preceding that of lactate), the rapid reaction studies showed that the overall steady state rate corresponded to that of NADH release (step ④ in the

---

† Faster processes than these can be monitored using relaxation techniques (Chapter 9) in which a system already at equilibrium is perturbed (e.g. by a rapid rise in temperature). From the rate at which the system relaxes to the new position of equilibrium the various rate constants in the mechanism can be deduced. By this type of method for instance it has been shown that the second order rate constant for association of malate dehydrogenase with NADH is approximately $7 \times 10^8 \ (mol \ dm^{-3})^{-1} s^{-1}$.

scheme below)

$$E + NAD^+ \rightarrow E^{NAD^+} \xrightarrow[\textcircled{1}]{\text{lactate}} E^{NAD^+}_{lactate}$$

$$\uparrow \qquad\qquad\qquad \textcircled{2} \downarrow$$

$$E \xleftarrow[\text{NADH}]{\textcircled{4}} E^{NADH} \xleftarrow[\text{pyruvate}]{\textcircled{3}} E^{NADH}_{pyruvate}$$

The rates of the elementary steps in the reaction (e.g. steps $\textcircled{1}$, $\textcircled{2}$, $\textcircled{3}$, and $\textcircled{4}$) can also be deduced affording a more complete description of the kinetics of the enzyme catalysed reaction.

(ii) In a reaction scheme such as that shown above where an intermediate (in this case E^{NADH}) is formed rapidly but breaks down only slowly then initially this intermediate will accumulate. If the breakdown of this intermediate is very slow indeed then a measure of the amount of intermediate accumulated (or other product released as the intermediate is formed) will give a measure of the amount of enzyme undergoing the reaction. This data can be used to evaluate the purity of an enzyme preparation since in effect we are determining the concentration of 'active sites'.

Additional studies on enzyme mechanisms

The studies of steady-state and pre-steady-state kinetics allow us to deduce which enzyme-containing complexes are kinetically significant in the overall reaction pathway and to assign rate constants to some or all of the elementary steps involved. However, if we are going to describe the mechanism completely we need to determine the structure of these complexes and to evaluate what parts of the enzyme molecule are involved in the binding of substrates and the catalytic processes. To answer these questions data from techniques such as X-ray crystallography (which can afford detailed structures of an enzyme and its complexes with substrates and substrate analogues) and chemical modification of amino acid side chains (which can determine which groups on the enzyme are important for the catalysis) are invaluable. Another useful technique in this respect is the study of the variation of enzyme activity with pH since this can, in favourable cases, give a good indication of the importance of certain amino-acid side chains in an enzyme.

The effect of pH on enzyme catalysed reactions

Changes in pH can have a number of effects on enzyme catalysed reactions. There could be (i) unfolding and consequent inactivation of the protein outside a certain pH range, (ii) changes in the position of equilibrium of a reaction if H^+ appears as a reactant or product in the overall equation, or (iii) ionization of groups in the substrate(s). These possibilities could be checked by performing appropriate experiments. However, we shall confine ourselves to a fourth possibility, namely changes in the ionisation state of amino acid side chains in the enzyme. The equations derived for such processes are similar to those used in Chapter 9 to describe the variation in reaction rate with pH (see p. 158).

The parameters K_m and V_{max} can both change with pH and a full kinetic analysis is necessary at each pH value (to check, for instance, that the substrate is still saturating). It is easier to discuss the variation of V_{max} with pH, since this generally reflects the variation of a single rate constant, whereas K_m is usually a function of several rate constants. (Of course it is possible that the rate determining step could be different at different pH values and this would complicate the analysis considerably.)

If a plot of $\log_{10} V_{max}$ for an enzyme catalysed reaction against pH is of the form shown in Fig. 10.17, it can be concluded that only the acidic form of the ionizing species is catalytically active. A pK_a for this group can be derived, as before, by extrapolating the linear portions of the curve (shown as ↓ in Fig. 10.17).

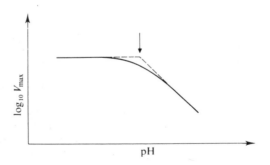

FIG. 10.17. Schematic representation of the effect of pH on the velocity of an enzyme-catalysed reaction if only the protonated form of an ionizing group is catalytically active. The solid curve represents the experimental data.

It is tempting to try to assign this observed pK_a to a particular type of amino acid side chain, and hence to implicate this amino acid as being involved in the mechanism of action of the enzyme. However, there are difficulties in this procedure, since the pK_as of amino acid side chains in enzymes might well be very different from the pK_a for the free amino acid. (An example of this is seen in the case of pepsin where an aspartic acid group involved in the mechanism has a pK_a of $1\cdot1$ compared with the 'normal' pK_a of the side chain of free aspartic acid (approximately 4·0). A very rough guide to the pK_a of some types of amino acid side chains in proteins is given in Table 10.1.

TABLE 10.1
Ionization properties of some amino acid side chains

Group	pK_a(298 K)	ΔH_i(kJ mol^{-1})
β-Carboxyl (Asp) $\}$ γ-Carboxyl (Glu) $\}$	~4	~±4
Imidazole (His)	~6	~29
ε-Amino (Lys)	~10	~46
Phenolic OH (Tyr)	~10	~25

Also included are some data on the enthalpies of ionization (ΔH_i) of these various groups, and it is possible that this could help in the assignment of pK_a values to specific groups (if the rate dependence on pH were known as a function of temperature).

Sometimes a second ionizing group may be present in the enzyme such as in the following scheme:

$$\underset{\underset{\text{inactive}}{E}}{\overset{\text{HX} \qquad \text{YH}}{\diagdown \diagup}} \quad \underset{pK_{a1}}{\overset{-\text{H}^+}{\underset{+\text{H}^+}{\rightleftharpoons}}} \quad \underset{\underset{\text{active}}{E}}{\overset{\text{HX} \qquad \text{Y}^-}{\diagdown \diagup}} \quad \underset{pK_{a2}}{\overset{-\text{H}^+}{\underset{+\text{H}^+}{\rightleftharpoons}}} \quad \underset{\underset{\text{inactive}}{E}}{\overset{\text{X}^- \qquad \text{Y}^-}{\diagdown \diagup}}$$

The plot of $\log_{10}(V_{max})$ against pH is then a curve of the type shown in Fig. 10.18.†

Again the values of pK_{a1} and pK_{a2} can be obtained by extrapolation of the appropriate linear portions of the plot (although if the two pK_as are close, say within 1·5 pH units, the ionizations will not be independent of each other and the pK_a values obtained by such procedures may need

---

† The equation is:

$$V_{max} = \frac{\bar{V}_{max}}{(1 + [\text{H}^+]/K_1 + K_2/[\text{H}^+])}$$

where $\bar{V}_{max}$ is the maximum velocity if all the enzyme is present in the active form, i.e.

$$\underset{E}{\overset{\text{XH}}{\diagup}}\diagdown_{\text{Y}^-}$$

The regions of the logarithmic plot (Fig. 10.18) of slope 1, 0, −1 correspond to the cases: (i) $[\text{H}^+] \gg K_1$; (ii) $K_1 \gg [\text{H}^+] \gg K_2$; (iii) $K_2 \gg [\text{H}^+]$ respectively.

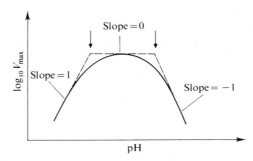

FIG. 10.18. The effect of pH on the velocity of an enzyme catalysed reaction when two ionizing groups are involved. The solid line is the experimental data.

correction to obtain the true pK_a values). An example of this type of analysis is afforded by fumarase (in the fumarate $\rightarrow$ malate direction) in the example below.

Worked example

(a) The data listed were obtain for the variation with pH of V_{max} for the reaction catalysed by fumarase at 298 K.

pH	5·2	5·5	6·0	6·5	6·75	7·0ᵃ	7·5	8·0	8·5
V_{max} (arbitrary units)	54	105	184	224	236	230	162	86	33

Comment on this data and deduce appropriate pK_a values.

Solution

The plot of $\log_{10} V_{max}$ against pH (Fig. 10.19) suggests that two ionizing groups are involved in the catalytic activity of fumarase. From the points of

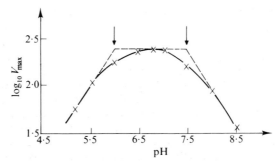

FIG. 10.19. Plot of $\log_{10} V_{max}$ against pH for the fumarase reaction.

intersection of the linear portions the pK_as can be determined as 5·9 and 7·5 for pK_{a1} and pK_{a2} respectively.

(b) It was found that at 308 K the lower pK_a was the same as at 298 K within 0·1 unit. What is the enthalpy of ionization of this group?

Solution

Since the pK_a is less than 0·1 unit different,

$$\log_{10}\left(\frac{K_{308}}{K_{298}}\right) \le 0\cdot 1$$

(where K refers to the ionization constant of the group of lower pK_a)

$$\therefore \quad \ln\left(\frac{K_{308}}{K_{298}}\right) \le 0\cdot 2303.$$

Now from eqn (3.9)

$$\ln\left(\frac{K_{T_2}}{K_{T_1}}\right) = -\frac{\Delta H^\circ}{R}\left(\frac{1}{T_2}-\frac{1}{T_1}\right)$$

we deduce that

$$-\frac{\Delta H^\circ}{R}\left(\frac{1}{308}-\frac{1}{298}\right) \le 0\cdot 2303,$$

i.e. $\underline{\Delta H^\circ \le 17\cdot 5 \text{ kJ mol}^{-1}}.$

The enthalpy of ionization of the group is thus 17·5 kJ mol^{-1} or less.

These data suggest that the group involved in the ionization may be a carboxyl group even though the pK_a value (5·9) is rather larger than normal. An imidazole group would be expected to possess a heat of ionization of about 29 kJ mol^{-1} (see Table 10.1).

The effect of temperature on enzyme-catalysed reactions

Many enzyme-catalysed reactions follow the normal Arrhenius equation for dependence of reaction rate against temperature. An apparent activation energy for the reaction can thus be derived. This activation energy would be expected to be lower than that for the non-catalysed reaction (see Chapter 9, p. 133). It has been shown that the overall rate equation has contributions from the various individual rate processes (characterized by k_1, k_{-1}, k_2, etc.) all of which we would expect to be temperature dependent. A detailed analysis of the system (such as studying the effect of temperature on the rate constants in the $E + S \rightleftharpoons ES$ equilibrium) would be required to derive the values of the activation energies of the various steps in the reaction.

We should note that there are several complications in the study of enzyme catalysed reactions. These include:

(i) Above a certain temperature, the enzyme molecule will become unfolded, so that the three-dimensional integrity of the catalytic site will become lost. The rate at which this process occurs is generally dependent upon the pH, the concentration of substrates or other ligands, ionic strength, etc. For many enzymes incubation at temperatures above about 323 K leads to fairly rapid denaturation, but some enzymes (e.g. those from thermophilic bacteria) remain stable at much higher temperatures. Generally, substrates protect an enzyme to some extent against heat inactivation.

In those cases where the rate of inactivation has been studied as a function of temperature, it has been found that the $\Delta S^{\circ\ddagger}$ for this process is normally very large and positive, as would be expected if the compact, active, structure of the enzyme has become substantially unfolded.

(ii) In a series of consecutive reactions (such as a multistep enzyme-catalysed reaction) with different activation energies, there can be a change in the rate-limiting step of the reaction with temperature. The rate of the reaction with the lower activation energy is less sensitive to temperature, and at a high enough temperature this will be the slow step of the reaction. This situation will result in a change in the slope of the Arrhenius plot at a certain temperature.

(iii) The enzyme may exist in two interconvertible, active forms which possess different activation energies. We would then expect a break in the Arrhenius plot around the temperature where the change-over between the two forms becomes significant (Fig. 10.20).

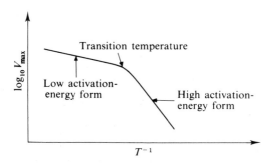

Fig. 10.20. Arrhenius plot for an enzyme-catalysed reaction, where the enzyme exists in two inter-convertible forms.

Two examples of this type of behaviour are seen with the ATPase enzymes which are involved with the transport of (a) Na^+ and K^+ and (b) Ca^{2+}. In these cases the transition probably arises from structural changes within the tightly bound phospholipid molecules which are associated with these enzymes. The possibility of structural changes in enzymes around the transition temperature could be checked by various means (e.g. sedimentation data, circular dichroism, fluorescence spectra, etc.).

Summary

This brief outline has shown how some of the important features of enzyme-catalysed reactions can be understood in terms of the relatively simple ideas used in a study of chemical processes. Clearly, a combination of techniques such as kinetics, X-ray crystallography, chemical modification, and spectroscopy is needed before a complete understanding of enzyme action is possible.

PROBLEMS

1. What are the effects on the Hanes plot ($[S]/v$ vs. $[S]$) of competitive, non-competitive, and uncompetitive inhibitors?

2. The liver contains two enzymes which can convert glucose to glucose-6-phosphate at the expense of ATP. One of these (hexokinase) has a K_m for glucose of 40 μmol dm^{-3} and a maximum catalytic activity of 0·7 μmol glucose transformed min^{-1} (g tissue)$^{-1}$. The other (glucokinase) has a K_m for glucose of 10 mmol dm^{-3} and a maximum catalytic activity of 4·3 μmol min^{-1} (g tissue)$^{-1}$. (The data refer to the enzymes from rat liver.)

   Use these results to calculate the proportion of glucose which is phosphorylated by glucokinase (as compared with hexokinase) in a fasting rat (glucose concentration = 3 mmol dm^{-3}) and in a well-fed rat (glucose concentration = 9·5 mmol dm^{-3}). Assume that ATP is present in saturating concentrations in each case.

3. Fructose diphosphatase (which catalyses the hydrolysis of fructose diphosphate to fructose-6-phosphate and phosphate) is inhibited by AMP. The following data were obtained from a study of the rat liver enzyme:

	[FDP] (μmol dm^{-3})				
[AMP] (μmol dm^{-3})	4	6	10	20	40
			Velocity (katal kg^{-1})		
0	0·059	0·076	0·101	0·125	0·150
8	0·034	0·043	0·056	0·071	0·083

Comment on these data.

4. The enzyme nucleoside diphosphokinase will catalyse the following reaction:

$$\text{GTP} + \text{dGDP} \xrightarrow{\text{Mg}^{2+}} \text{GDP} + \text{dGTP}.$$

In an experiment with the enzyme isolated from erythrocytes, the following results were obtained:

	[GTP] (μmol dm^{-3})			
	22	30	50	200
[dGDP] (μmol dm^{-3})		Velocity (katal kg^{-1})		
20	0·095	0·112	0·141	0·196
25	0·102	0·120	0·155	0·223
40	0·112	0·136	0·180	0·284
100	0·125	0·156	0·218	0·385

What can you deduce from these data regarding a likely mechanism for the enzyme catalysed reaction? How could you check your conclusion?

5. A study was made of the reaction catalysed by rabbit muscle creatine kinase:

$$\text{creatine} + \text{ATP} \xrightarrow{\text{Mg}^{2+}} \text{phosphocreatine} + \text{ADP}.$$

The following data were obtained:

	[ATP] (mmol dm^{-3})			
	0·46	0·62	1·23	3·68
[Creatine] (mmol dm^{-3})		Velocity (katal kg^{-1})		
6	0·377	0·463	0·660	0·968
10	0·555	0·678	0·950	1·308
20	0·845	1·005	1·338	1·803
40	1·180	1·378	1·718	2·295

Evaluate the kinetic parameters for this reaction. From other data it appears that the reaction proceeds via a random order ternary complex mechanism. What can you deduce about the binding of substrates to the enzyme?

6. The effect of one of the products (pyruvate) on the reaction catalysed by rabbit muscle lactate dehydrogenase was studied, with the following results:

$$\text{NAD}^+ + \text{lactate} \rightarrow \text{NADH} + \text{pyruvate}.$$

At a fixed NAD^+ concentration (1.5 mmol dm^{-3})

[Pyruvate] (μmol dm^{-3})	[Lactate] (mmol dm^{-3})			
	1·5	2·0	3·0	10·0
	Velocity (katal kg^{-1})			
0	1·88	2·36	3·10	5·81
40	1·05	1·34	1·88	4·19
80	0·73	0·94	1·34	3·27

At a fixed lactate concentration (15 mmol dm^{-3})

[Pyruvate] (μmol dm^{-3})	[NAD^+] (mmol dm^{-3})			
	0·5	0·7	1·0	2·0
	Velocity (katal kg^{-1})			
0	3·33	3·91	4·50	5·42
30	2·65	3·13	3·60	4·33
60	1·97	2·30	2·66	3·21

What types of inhibition are being observed in these cases?

7. The velocity of the reaction catalysed by pyruvate kinase was studied as a function of phosphoenolpyruvate (PEP) concentration with the results shown in row A.

$$\text{phosphoenolpyruvate} + \text{ADP} \xrightarrow{\text{Mg}^{2+}} \text{pyruyate} + \text{ATP}.$$

(It is assumed that ADP is present in saturating concentrations.)

Also given are data on the effects of two inhibitors of this reaction. In row B the data refer to 10 mmol dm^{-3} 2-phosphoglycerate as inhibitor, and in row C, the data refer to 10 mmol dm^{-3} phenylalanine as inhibitor. Comment on these data.

Inhibitor	[PEP] (mmol dm^{-3})				
	0·25	0·3	0·5	1·0	2·0
	Velocity (katal kg^{-1})				
A None	1·10	1·25	1·63	2·05	2·43
B 2-Phosphoglycerate	0·73	0·83	1·15	1·68	2·08
C Phenylalanine	0·60	0·68	0·90	1·25	1·50

8. The activity of the enzyme nitrogenase (which catalyses the reduction of nitrogen to ammonia) from *Azotobacter* was studied as a function of temperature with the following results:

T (K)	278	284	290	296	302	308	314	320	326	330
Velocity (arbitrary units)	1·65	11·7	72·2	284	483	781	1180	1260	200	9·5

Comment on these data.

9. The trypsin catalysed hydrolysis of N-benzoyl-L-arginine ethyl ester was studied (at 298 K) as a function of pH with the following results.

pH	4·0	4·5	5·0	5·5	6·0	6·5
V_{max} (katal kg^{-1})	0·0028	0·0087	0·027	0·076	0·180	0·320

pH	7·0	7·5	8·0	8·5	9·0
V_{max} (katal kg^{-1})	0·425	0·473	0·491	0·497	0·499

What is the pK_a of the ionizing group in this process? At 308 K the pK_a of this group is 6·08. What is the enthalpy of ionization of this group?

10. Chymotrypsin catalyses the hydrolysis of p-nitrophenylacetate (PNPA) to yield p-nitrophenol and acetate. The production of p-nitrophenol (PNP) was monitored as a function of time with the following results:

Experiment 1 [chymotrypsin] = 23 μmol dm^{-3} : [PNPA] = 500 μmol dm^{-3}

t(s)	0	15	30	60	90	120	150	180
[PNP] (μmol dm^{-3})	0	18·3	22·2	29·1	35·7	41·7	48·0	54·3

Experiment 2 [chymotrypsin] = 12 μmol dm^{-3} : [PNPA] = 500 μmol dm^{-3}

t(s)	0	15	30	60	90	120	150	180
[PNP] (μmol dm^{-3})	0	10·3	12·6	16·2	19·7	23·3	26·8	30·4

What can you deduce from these data?

11. Spectrophotometry

The electromagnetic spectrum

SPECTROPHOTOMETRY refers to the measurement of absorption of electromagnetic radiation (e.g. light) by compounds. The basic condition for absorption of radiation of a given frequency ν is that there are two *energy levels* of the compound separated by an amount of energy E where

$$E = h\nu$$

h is Planck's constant, $6 \cdot 63 \times 10^{-34}$ J s.

What are these energy levels? The results of quantum mechanics show that a molecule can possess different types of energy such as that associated with rotation of the molecule (or parts of the molecule) about certain axes, or that associated with vibration of atoms in the chemical bonds, etc. However, the type of energy with which we shall principally be concerned is known as *electronic energy*. This is determined by the distribution of electrons within the various available orbitals† of the molecule. Absorption of radiation promotes electrons from a given energy level to one of higher energy (Fig. 11.1).

FIG. 11.1. Absorption of radiation of frequency ν.

Worked example

What is the energy corresponding to radiation of wavelength 589 nm? (This is the so called sodium D line—commonly observed by putting sodium salts into flames.)

Solution

The frequency ν is related to wavelength λ by:

$$\nu = c/\lambda$$

where c is the velocity of light $= 3 \times 10^8$ m s^{-1}.

---

† The student unfamiliar with such terms should note that these concepts are not essential in order to understand the applications in this Chapter.

Thus

$$\nu = \frac{3 \times 10^8}{589 \times 10^{-9}} = 5 \cdot 1 \times 10^{14} \text{ Hz.}$$

(N.B. 1 nm $= 10^{-9}$ m) From the formula $E = h\nu$ we obtain

$$E = 6 \cdot 63 \times 10^{-34} \times 5 \cdot 1 \times 10^{14} \text{ J}$$

$$E = 3 \cdot 38 \times 10^{-19} \text{ J}$$

This is the energy associated with *one* atom. If we wished to express it as an energy per mole, we must multiply this number by Avogadro's number $(6 \cdot 03 \times 10^{23})$, whence

$$E = 2 \cdot 04 \times 10^5 \text{ J mol}^{-1}$$

$$E = 204 \text{ kJ mol}^{-1}.$$

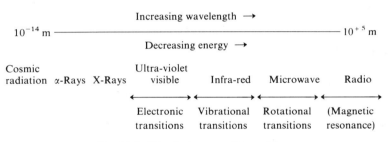

FIG. 11.2. The electromagnetic spectrum.

The complete electromagnetic spectrum is represented in Fig. 11.2. A *spectrum* usually consists of a plot of the absorption of radiation as a function of the wavelength or frequency. So far we have dealt with the wavelength (i.e. energy) of absorption. We now consider the amount of absorption. This is described by the *Beer–Lambert Law*.

The Beer–Lambert Law states that

$$\log_{10}\left(\frac{I_0}{I_t}\right) = \varepsilon c l$$

where I_0 is the radiation incident on the compound, I_t is the radiation transmitted by the compound (i.e. $I_0 - I_t$ is the radiation absorbed), c is the concentration of the compound, and l is the length of cell through which radiation actually travels (i.e. the path length). ε is known as the *extinction coefficient* of the compound at the particular wavelength of the incident radiation and is thus a measure of the absorbing power of the compound. The units of ε will depend on those chosen for c and l. The term $\log_{10}(I_0/I_t)$ is known as either the *absorbance* (A) or *optical density* (O.D.) of the compound and is clearly dimensionless and hence the units of the product (εcl) must also be dimensionless.

Worked examples

(1) 15·8 per cent of the 340 nm radiation incident on a certain solution of NADH is transmitted. Given that the extinction coefficient of NADH at this wavelength is $6·22 \times 10^6$ cm² mol⁻¹,† what is the concentration of NADH in the solution? The path length is 1 cm.

Solution

If

$$I_t = \frac{15·8}{100} I_0$$

the *absorbance* (A) is given by

$$A = \log_{10}\left(\frac{I_0}{I_t}\right)$$

$$= 0·801.$$

Now

$$A = \varepsilon cl$$

and

$$\varepsilon = 6·22 \times 10^6 \text{ cm}^2 \text{ mol}^{-1},$$

$$l = 1 \text{ cm}.$$

Hence

$$c = \frac{0·801}{6·22 \times 10^6} \text{mol cm}^{-3}$$

$$= \frac{0·801}{6·22 \times 10^3} \text{mol dm}^{-3}$$

$$\underline{c = 1·29 \times 10^{-4} \text{ mol dm}^{-3}}$$

---

† The extinction coefficient is sometimes expressed as $6·22 \times 10^3$ (mol dm⁻³)⁻¹ cm⁻¹, which is of course entirely equivalent. It should be noted that we have retained cm (a supplementary SI unit) for the purposes of convenience in this section.

Since the units of ε were given as $cm^2\,mol^{-1}$ and that of l was cm, the concentration is expressed as $mol\,cm^{-3}$. This is converted to $mol\,dm^{-3}$ by multiplication by 10^3.

(2) Calculate the absorbance at 340 nm of (a) a 1 mmol dm^{-3} solution and (b) a 1 μmol dm^{-3} solution of NADH in a cell of path length 1 cm. Calculate the percentage transmitted radiation in these two cases.

Solution

From the Beer–Lambert Law

$$A = \varepsilon cl.$$

ε is in units of $cm^2\,mol^{-1}$, so c is expressed in $mol\,cm^{-3}$.

Case (a)

$$A = (6\cdot22 \times 10^6)(1 \times 10^{-3} \times 10^{-3})(1)$$

i.e. $\underline{A = 6\cdot22}$

$$\log_{10}\left(\frac{I_0}{I_t}\right) = 6\cdot22$$

i.e. $\underline{I_t = 6\cdot01 \times 10^{-5}\% I_0.}$

Case (b)

$$A = (6\cdot22 \times 10^6)(1 \times 10^{-6} \times 10^{-3})(1)$$

i.e. $\underline{A = 0\cdot00622.}$

In this case

$$\underline{I_t = 98\cdot6\% I_0}$$

Appropriate concentration ranges

Clearly it would be extremely difficult to measure the concentration of either of the solutions in the worked example accurately by spectrophotometry. In the first case the intensity of transmitted radiation is negligibly small, whereas in the second case the amount of radiation actually absorbed is very small. In practice I_t should be between 10 per cent and 90 per cent of I_0 (i.e. the absorbance should be between $1\cdot0$ and $0\cdot05$), but the most accurate measurements will be where $I_t \sim 50$ per cent of I_0 (i.e. absorbance $= 0\cdot3$). These conditions can be realized by alteration of the concentration of the absorbing compound, the sample path length or by changing the wavelength of the radiation (i.e. by changing ε). The first procedure is most commonly employed.

Worked example

By what factor would you dilute 1 mmol dm^{-3} solution of NADH to measure its absorbance at 340 nm accurately? The path length can be assumed to be 1 cm.

Solution

The absorbance can be most accurately measured when it is approximately equal to 0·3.

Now $A = \varepsilon c l$.

For a path length of 1 cm, $A = 0\cdot3$ corresponds to a concentration of $4\cdot8 \times 10^{-5}$ mol dm^{-3} (i.e. 48μmol dm^{-3}), i.e. we must dilute the 1 mmol dm^{-3} solution 20 fold to determine the absorbance accurately.

If it is undesirable to dilute the solution, we could use short pathlength cells (e.g. 1 mm) and measure the absorption of radiation at a different wavelength, e.g. 366 nm (where ε is known to be $3\cdot3 \times 10^{6}$ cm^2 mol^{-1}).

The Beer–Lambert Law is very widely used to provide a simple means of estimating the concentration of an absorbing compound. Thus the concentrations of solutions of nucleotides and nucleic acids can be estimated by their absorption of 260 nm radiation. For proteins a suitable wavelength is 280 nm. Of course in each case we have to know the value of ε.

Two absorbing compounds

In many systems of interest there is more than one absorbing compound, and in some cases the spectra of these overlap (Fig. 11.3).

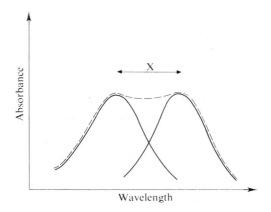

FIG. 11.3. Two overlapping absorption bands. The continuous lines are the individual spectra and the broken line represents the sum of the two (which is actually observed).

The Beer–Lambert Law is, of course, applicable to each compound separately. However in the region marked X in Fig. 11.3, the total absorbance contains contributions from both species. We can determine the concentrations of the individual compounds either by making measurements at a wavelength where there is no overlap, or by making measurements at two wavelengths and using simultaneous equations, as illustrated in the example below.

Worked example

The absorbance of a sample of (enzyme + AMP) is 0·46 at 280 nm and 0·58 at 260 nm. Calculate the concentration of each component given that for the enzyme

$$\varepsilon_{280} = 2\cdot96 \times 10^7 \text{ cm}^2 \text{ mol}^{-1} \quad \text{and} \quad \varepsilon_{260} = 1\cdot52 \times 10^7 \text{ cm}^2 \text{ mol}^{-1}$$

and for AMP

$$\varepsilon_{260} = 1\cdot5 \times 10^7 \text{ cm}^2 \text{ mol}^{-1} \quad \text{and} \quad \varepsilon_{280} = 2\cdot4 \times 10^6 \text{ cm}^2 \text{ mol}^{-1}.$$

The pathlength is 1 cm.

Solution

Let the concentrations of enzyme and AMP be x and y mmol dm^{-3} (i.e. μmol cm^{-3}) respectively.

Since $A = \varepsilon c l$ for each component and

(Absorbance)$_\lambda$ = (Absorbance of enzyme)$_\lambda$ + (Absorbance of AMP)$_\lambda$,

where λ is the given wavelength, we obtain

at 260 nm: $0\cdot58 = 15\cdot2x + 15y$

and at 280 nm: $0\cdot46 = 29\cdot6x + 2\cdot4y.$

Solution of these two equations gives $x = 0\cdot0135$, $y = 0\cdot0252$, i.e. the concentrations of enzyme and AMP are 13·5 μmol dm^{-3} and 25·2 μmol dm^{-3} respectively.

Isosbestic points

A special case of overlapping spectra occurs when the two compounds concerned are in equilibrium with each other. Consider the ionization of *ortho*-nitrotyrosine:

The neutral molecule (I) and its anion (II) have maximum absorptions at 360 nm and 428 nm respectively. (The wavelength of maximum absorption is often abbreviated as λ_{max}.)

As we alter the relative contributions of the two compounds (in this case by changing the pH) there will be a shift in the maximum of the absorption spectrum, as illustrated in Fig. 11.4.

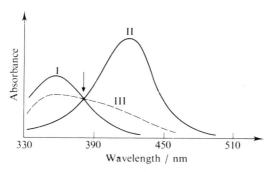

FIG. 11.4. Schematic representation of spectra of pure *o*-nitrotyrosine (I), its anion (II), and a mixture (III). Note the isosbestic point, indicated by the arrow.

Although the intensity of absorption varies in going from (I) to (II), there is one wavelength at which the intensity remains constant. This is known as an *isosbestic point*. If an isosbestic point is observed while some experimental parameter (e.g. pH) is varied in a titration, this means that *two* (and only two) absorbing species are in equilibrium with each other. Measurements of absorption at the isosbestic point are often used to obtain the *total* concentration of absorbing species. Of course, in order to obtain the individual concentrations, we must make measurements at two wavelengths as described previously.

PROBLEMS

1. What energy (in $kJ\,mol^{-1}$) does electromagnetic radiation of the following wavelengths represent?
   (*a*) 250 nm (typical value for ultraviolet (u.v.) absorption).
   (*b*) 5000 nm (typical value for infrared (i.r.) absorption).
   (*c*) 1 cm (typical value for electron spin resonance (e.s.r.) absorption).
   (*d*) 500 cm (typical value for nuclear magnetic resonance (n.m.r.) absorption).
   (*e*) 247 m (typical value for B.B.C.).

2. The human eye (when completely adapted to darkness) is able to perceive a point source of light against a dark background when radiation falls on the retina at a

rate of 2×10^{-16} J s^{-1}. What is the minimum rate of incidence of quanta of radiation (photons) on the retina which can be perceived, assuming an incident wavelength of 550 nm?

3. The absorbance at 280 nm of a 1 g dm^{-3} solution of adenylate kinase (M.W. = 21 000) is 0·53, in a 1 cm cell. Calculate ε, and state clearly the units.

4. The absorbance (in a cell of 1 cm path length) of a solution containing NAD$^+$ and NADH is 0·21 at 340 nm and 0·85 at 260 nm. The extinction coefficient of NAD$^+$ and of NADH at 260 nm is $1·8 \times 10^7$ cm^2 mol^{-1}, while the value at 340 nm for NADH is $6·22 \times 10^6$ cm^2 mol^{-1}. (NAD$^+$ does not absorb at 340 nm.)
   Calculate the concentrations of NAD$^+$ and NADH in the solution.

5. The ionization of *ortho*-nitrotyrosine produces a yellow colour:

λ_{max} 360 nm   428 nm

The following titration data were obtained for (*a*) nitrotyrosine and (*b*) the nitrotyrosine group in a sample of nitrated glutamate dehydrogenase.

(*a*)		(*b*)	
	Extinction coefficient of nitrotyrosine at 428 nm		Extinction coefficient of nitrated enzyme at 428 nm
pH	cm^2 mol^{-1} ($\times 10^{-5}$)	pH	cm^2 mol^{-1} ($\times 10^{-5}$)
4·6	1·6	5·6	2·0
5·4	3·6	6·4	3·8
6·0	7·0	7·2	7·6
6·6	16·0	7·8	16·0
7·0	27·6	8·2	28·0
7·5	37·0	8·8	34·6
8·6	41·0	9·6	37·6
10·0	42·0	10·8	39·0

Determine the pK_a values of the nitrotyrosine in these two cases, and suggest reasons why they might differ.

6. The enzyme creatine kinase (CK) is often assayed according to the following scheme:

$$\text{creatine} + \text{ATP} \xrightarrow{\text{CK}} \text{creatine} - Ⓟ + \text{ADP}$$

$$\text{ADP} + Ⓟ\text{-enolpyruvate} \xrightarrow[\text{kinase}]{\text{pyruvate}} \text{ATP} + \text{pyruvate}$$

$$\text{pyruvate} + \text{NADH} \xrightarrow[\text{dehydrogenase}]{\text{lactate}} \text{lactate} + \text{NAD}^+$$

The oxidation of NADH is monitored at 340 nm (where NAD^+ does not absorb). In a cell of path length 1 cm the absorbance of a 1 mmol dm^{-3} solution of NADH at 340 nm is 6·22.

(a) 0·02 cm^3 of a solution of creatine kinase whose concentration is 0.08 g dm^{-3} is added at 298 K to 1 cm^3 of assay mixture (containing all the appropriate substrates, cofactors and coupling enzymes in saturating amounts), in a cell of path length 1 cm. The decrease in absorbance at 340 nm is 0·52 per minute. What is the activity of the creatine kinase solution in μmol substrate consumed min^{-1} (mg enzyme)$^{-1}$. (These are known as International Units of enzyme activity.) What is this activity expressed as katal kg^{-1} (SI units; see Chapter 10).

(b) The above assay system can also be used to determine the concentration of a solution of creatine.

A stock solution of creatine was diluted 50 fold into an appropriate assay mixture. After addition of creatine kinase the absorbance at 340 nm had decreased by 0·8. Calculate the concentration of creatine in the stock solution. (It can be assumed that the creatine kinase equilibrium is driven well over to the right.)

7. In the reaction catalysed by glutamate dehydrogenase

$$\text{Glutamate}^- + NAD^+ + H_2O \rightleftharpoons \alpha\text{-ketoglutarate}^{2-} + NADH + NH_4^+ + H^+$$

the initial concentrations of glutamate and NAD^+ are 20 mmol dm^{-3} and 1 mmol dm^{-3} respectively. The reaction was carried out in phosphate buffer at pH = 7 and the concentration of NADH was measured spectrophotometrically. The absorbance at 340 nm (in a cell of path length 1 cm) increased until a constant value of 0·561 was reached. Calculate the equilibrium constant for the above reaction. (ε for NADH = $6·22 \times 10^6$ cm^2 mol^{-1} at 340 nm). Comment on the major sources of error.

8. The haemoglobin (Hb) content of a solution can be determined by treating the solution with a reagent solution containing an excess of ferricyanide and cyanide. This converts both Hb and oxy-Hb to cyanomet-Hb which can be determined from its absorbance at 540 nm. The following results were obtained with a standard haemoglobin solution (0·6 g dm^{-3}):

Volume (cm^3) of standard Hb solution added to reagent (total volume = 5 cm^3)	Absorbance at 540 nm
0	0·025
1	0·090
2	0·160
3	0·230
4	0·290

When 0·01 cm^3 of a blood sample was added to 5 cm^3 of the reagent solution, the absorbance at 540 nm was 0·19. What is the concentration of Hb in the blood sample?

12. Isotopes in biochemistry

The uses of isotopes

THERE are two main types of uses of radioactive isotopes in biochemistry.

(1) Analytical uses

Radioactivity provides an extremely sensitive method of analysis since disintegrations of individual nuclei are monitored. If a radioactive atom (e.g. ^{3}H or ^{14}C) can be incorporated into a given compound, then we have a means of assaying this compound under a variety of conditions simply by measuring its radioactivity. This is utilized in the worked example below and in the technique of 'isotopic dilution' (Problem 1).

Worked example. 100 mg of enzyme (mol wt 50 000) were reacted with excess ^{14}C-labelled iodoacetic acid (whose radioactivity† was 60 Ci mol^{-1}). After dialysis to remove the unreacted iodoacetic acid, 0·1 mg of the enzyme was taken for analysis. This sample gave $5·05 \times 10^5$ disintegrations per minute (d.p.m.). How many groups on the enzyme were labelled? (1 Curie (Ci) is equivalent to $3·7 \times 10^{10}$ disintegrations per second, d.p.s.).

Solution

$$100 \text{ mg of the enzyme} = \frac{100 \times 10^{-3}}{50\,000} \text{ mol}$$

$$= 2 \times 10^{-6} \text{ mol}.$$

For the analysis we take 0·1 mg (i.e. 2×10^{-9} mol).

The radioactivity of this sample $= 5·05 \times 10^5$ d.p.m.

$$= \frac{5·05 \times 10^5}{60} \text{ d.p.s.}$$

Now since 1 Curie $= 3·7 \times 10^{10}$ d.p.s. the radioactivity is

$$\frac{5·05}{60} \times \frac{10^5}{3·7 \times 10^{10}} \text{ Ci}$$

i.e. $\underline{2·27 \times 10^{-7} \text{ Ci}}$.

We can convert this radioactivity to moles by dividing by 60 (since the iodoacetic acid has a specific activity of 60 Ci mol^{-1}).

---

† This is often referred to as the *specific activity* of the compound, in the appropriate units. Strictly speaking in SI units, the Ci is redundant $1 \text{ Ci} = 3·7 \times 10^{10} \text{ s}^{-1}$.

Thus the radioactivity incorporated is equivalent to

$$\frac{2 \cdot 27 \times 10^{-7}}{60} \text{ mol} = \underline{3 \cdot 78 \times 10^{-9} \text{ mol}.}$$

Since we have 2×10^{-9} mol of enzyme in this sample, the extent of incorporation is 1·9 *mol iodoacetic acid per mol of enzyme.*

This example illustrates the great sensitivity of the radioactivity method. In practice it would be possible to measure radioactivity one thousand fold less than this (i.e. from 10^{-4} mg enzyme).

If the radioactive compound undergoes subsequent reaction in the system (it might be, for instance, an intermediate in a biochemical pathway), we can use the radioactivity to follow the fate of a particular atom of the compound in these pathways.

The analytical use of radioactivity can be extended by incorporation of two different radioactive atoms in a compound. For instance ^{3}H and ^{14}C emit electrons (β-particles) with very different kinetic energies, as shown in Fig. 12.1.

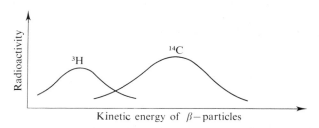

FIG. 12.1. Schematic representation of the distribution of the energies of β-particles emitted from the isotopes ^{3}H and ^{14}C.

Radioactivity is normally measured across a *range* of kinetic energies (compare an absorption spectrum of a compound), and so by making observations at appropriate ranges of energy, it is possible to determine the individual contributions from the two isotopes.

(2) *Kinetic uses*

Radioactive decay is a first-order process. In Chapter 9 (eqn. (9.1)) we noted that the equation for such a process was

$$\ln([A]/[A_0]) = -kt.$$

In this case we use this symbol N_t for the number of radioactive atoms remaining after time t;

i.e. $\boxed{\ln(N_t/N_0) = -kt}$,

where k is the *rate constant* (decay constant) and the half life of the isotope $(t_\frac{1}{2})$ is given by

$$\boxed{t_\frac{1}{2} = \frac{\ln 2}{k} = \frac{0 \cdot 693}{k}}.$$

Sometimes other rate processes contribute to the overall decay of radioactivity. One example would be when a radioactive compound is injected into an animal. The subsequent decline in radioactivity of (say) a plasma sample with time results not only from the radioactive decay of the isotope but also because some of the injected compound will have been excreted by the animal. In practice this latter process appears to be first order in many cases, and is assigned a half life known as the *biological* half life.

PROBLEMS

1. The technique of 'isotopic dilution' is often used in analytical work. 10 microcuries (μCi) of a sample of ^{14}C-phenylalanine (of specific activity of 50 Ci mol^{-1}) was added to a mixture of amino acids. A small sample of the phenylalanine present was then isolated and found to possess an activity of 2000 disintegrations per minute (d.p.m.) per mg.
   How much phenylalanine was present in the mixture of amino acids? (1 Ci is equal to $3 \cdot 7 \times 10^{10}$ disintegrations per second.)

2. One advantage of radioactive methods is their great sensitivity. This is illustrated in the following application. ^{32}P is an artificially-produced β-emitter and has a half-life of 14·3 days. A specimen of rock suspected of containing phosphorus (mainly as potassium phosphate) was treated to convert all the phosphorus atoms to ^{32}P. The rate of emission of β-particles from this sample was 973 disintegrations min^{-1}. Calculate the number of atoms of phosphorus in the rock.

3. A sample containing 10 μg tRNA were treated with 1 mCi ^{14}C-L-alanine (of specific activity 90 Ci mol^{-1}), in the presence of the appropriate enzyme and cofactors. The L-alanine–tRNA was then separated and its radioactivity was found to be 60 000 d.p.m. Calculate the percentage of the tRNA in the form of L-alanine-tRNA.
   (1 Ci is equal to $3 \cdot 7 \times 10^{10}$ disintegrations per second. The molecular weight of tRNA is 25 000.)

4. What is meant by the decay constant and the half-life of an isotope?
   A compound X labelled with an isotope having a decay constant of 0·08 day^{-1} was injected into rats, and the radioactivity of plasma samples was determined at different times. The results were as follows:

Time (days)	2	6	10	16
Radioactivity (d.p.m. cm^{-3}) $\times 10^{-3}$	68·9	32·6	15·64	5·27

What is the biological half-life of X?

Appendix 1

The dependence of enthalpy and entropy on pressure and temperature

Enthalpy

CONSIDER a reaction in which a moles A and b moles B are converted into l moles L and m moles of M, i.e.

$$aA + bB \rightarrow lL + mM$$

The enthalpy of the products H_{final} is given by:

$$H_{\text{final}} = l(Hm)_L + m(Hm)_M,$$

where $(Hm)_L$ and $(Hm)_M$ are the molar enthalpies of L and M. Similarly

$$H_{\text{initial}} = a(Hm)_A + b(Hm)_B,$$

ΔH for the reaction is thus

$$\Delta H = H_{\text{final}} - H_{\text{initial}}.$$

The variation of ΔH with temperature is given by

$$\frac{d}{dT}(\Delta H) = \frac{d}{dT}(H_{\text{final}}) - \frac{d}{dT}(H_{\text{initial}}).$$

But

$$\frac{dHm}{dT} = C_p \text{ (the molar specific heat)†.}$$

Thus:

$$\frac{d(\Delta H)}{dT} = l(C_p)_L + m(C_p)_M - a(C_p)_A - b(C_p)_B.$$

which is written in a shorthand notation as:

$$\boxed{\frac{d(\Delta H)}{dT} = \Delta C_p}.$$

This is known as *Kirchoff's Law*.

If we consider the reaction carried out at two temperatures T_1 and T_2 we can integrate the above equation viz.:

$$\int_{T_1}^{T_2} \frac{d(\Delta H)}{dT} = \int_{T_1}^{T_2} \Delta C_p,$$

---

† C_p is the molar specific heat and is defined as the amount of heat at constant pressure, which one mole must absorb to raise its temperature by 1 K.

or

$$\int_{T_1}^{T_2} d(\Delta H) = \int_{T_1}^{T_2} \Delta C_p \, dT.$$

Thus

$$\Delta H_{(T_2)} - \Delta H_{(T_1)} = \Delta C_p \,(T_2 - T_1),$$

or

$$\Delta H_{(T_2)} = \Delta H_{(T_1)} + \Delta C_p (T_2 - T_1).$$

We should note that the change in ΔH with temperature is not very marked (especially in the range of temperatures of biological interest), and thus ΔH is taken to be independent of temperature. A significant exception occurs in the case of proteins. Thus for the denaturation of lysozyme by guanidinium hydrochloride, ΔH^0 is 90·3 kJ mol^{-1} at 298 K and $\Delta C_p = 5 \cdot 5$ kJ K^{-1} mol^{-1}. Thus ΔH^0 at 293 K is 62·7 kJ mol^{-1} and at 303 K is 117·9 kJ mol^{-1}. The large value of ΔC_p arises because of the break up of the three-dimensional structure on denaturation—which involves the rupture of many non-covalent bonds.

For ideal solutions or ideal gases, ΔH is independent of pressure.

Entropy

The starting point here is the equation

$$H = U + PV.$$

Thus

$$dH = dU + P \, dV + V \, dP.$$

Substituting for dU from the first law of thermodynamics

$$dH = dq - P \, dV + P dV + V \, dP.$$

Using the second law $dq = T \, dS$, so that

$$dH = T \, dS + V \, dP$$

$$\therefore \ dS = \frac{dH}{T} - \frac{V \, dP}{T}.$$

Since

$$\frac{dH}{dT} = nC_p \text{ (for } n \text{ moles)},$$

(A.1.1)
$$\therefore \ \boxed{dS = \frac{nC_p \, dT}{T} - \frac{V \, dP}{T}}.$$

Consider a process in which the temperature changes from T_1 to T_2 at *constant pressure* $(dP = 0)$. Then

$$\Delta S = S_2 - S_1 = \int_{T_1}^{T_2} nC_p \, dT/T,$$

$$\therefore \ \underline{\Delta S = nC_p \ln(T_2/T_1)},$$

(assuming that C_p is independent of temperature). Thus the increase in entropy on heating one mole of water from 273·15 K to 373·15 K, where $C_p = 75\cdot2\ \mathrm{J\ mol^{-1}\ K^{-1}}$ is 23·4 J K^{-1} mol^{-1}.

The absolute entropy of a substance at any temperature can be determined using the equation

$$\Delta S = nC_p \ln(T_2/T_1),$$

and the Third Law of Thermodynamics, which states that 'at absolute zero the entropy of all substances becomes zero, i.e. they become perfectly ordered.' We can determine absolute entropies by making measurements of C_p at various temperatures (down towards 0 K) and using the integrated form of the above equation to determine S at the temperature in question. If there are phase changes, the extra entropy change ($\mathrm{d}S = \mathrm{d}q/T$) must be included.

Thus a graph of S vs. T would generally be of the form shown in Fig. A.1.

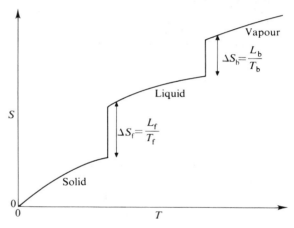

Fig A.1. Variation of entropy with temperature.

If we now investigate the process at *constant temperature* in which the pressure changes from P_1 to P_2 then ($\mathrm{d}T = 0$) and from eqn (A.1.1).

$$\Delta S = S_2 - S_1 = -\int_{P_1}^{P_2} \frac{V\ \mathrm{d}P}{T}$$

which, for instance, for an ideal gas would give

$$\Delta S = nR \ln(P_1/P_2).$$

A similar expression could be derived for an ideal solution by using Raoult's law. We can use this result to consider any change in entropy on mixing.

Consider the change in entropy on mixing *one* mole of oxygen (1 atm.) with *four* moles of nitrogen to give air at 1 atm. The final pressure of the oxygen (i.e. partial pressure) is 1/5 atm. Thus ΔS for oxygen is $R \ln 5 = 13\cdot4\ \mathrm{J\ K^{-1}\ mol^{-1}}$. ΔS for nitrogen is $4R \ln(1/\frac{4}{5}) = 7 \cdot 4$ J K^{-1} mol^{-1}. (Note that $n = 4$.) The total entropy increase of mixing is **20·8** J K^{-1} mol^{-1}.

If the n_{N_2} moles of nitrogen originally at pressure P is mixed with O_2 such that the final pressure is P, the partial pressure of N_2 is $(X_{N_2} \cdot P)$ where X_{N_2} is the mole fraction of N_2 in the mixture. We then have from above

$$(\Delta S_{\text{mixing}})_{N_2} = n_{N_2} \cdot R \ln(P/X_{N_2}P),$$

i.e.

$$(\Delta S_{\text{mixing}})_{N_2} = -n_{N_2} \cdot R \ln X_{N_2},$$

or, in general, for i components

$$\Delta S_{\text{mixing}} = -R \sum_i n_i \ln X_i.$$

This result is, of course, applicable to either ideal solutions or ideal gases.

Appendix 2

The equation for multiple binding sites

CONSIDER the system in which one mole of macromolecule P can bind up to n moles of ligand, A.

		Sites occupied	Sites unoccupied
$P+A \rightleftharpoons PA$		1	$n-1$
$PA+A \rightleftharpoons PA_2$		2	$n-2$
$PA_2+A \rightleftharpoons PA_3$		3	$n-3$
$PA_{n-1}+A \rightleftharpoons PA_n$		n	0

Let the successive dissociation constants for PA, PA_2 ... be K_1, K_2 etc.

i.e. $$K_1 = \frac{[P][A]}{[PA]}, \quad K_2 = \frac{[PA][A]}{[PA_2]} \text{ etc.}$$

Now the average number of moles of A bound per mole P, (r) is given by

$$r = \frac{[\text{total concentration of A bound}]}{[\text{total concentration of P}]}$$

$$= \frac{[PA]+2[PA_2]+3[PA_3]+\ldots}{[P]+[PA]+[PA_2]+[PA_3]+\ldots} \quad †$$

Now

$$[PA] = \frac{[P][A]}{K_1}$$

and

$$[PA_2] = \frac{[PA][A]}{K_2}$$

$$= \frac{[P][A]^2}{K_1 K_2}$$

and

$$[PA_3] = \frac{[P][A]^3}{K_1 K_2 K_3} \text{ etc.}$$

$$\therefore \quad r = \frac{\dfrac{[P][A]}{K_1} + \dfrac{2[P][A]^2}{K_1 K_2} + \dfrac{3[P][A]^3}{K_1 K_2 K_3} + \ldots}{[P] + \dfrac{[P][A]}{K_1} + \dfrac{[P][A]^2}{K_1 K_2} + \dfrac{[P][A]^3}{K_1 K_2 K_3} + \ldots}$$

† Note the factors 1, 2, 3, etc. in the numerator. These arise because each mole of PA_n contains n moles of A.

$$= \cfrac{\dfrac{[A]}{K_1} + \dfrac{2[A]^2}{K_1K_2} + \dfrac{3[A]^3}{K_1K_2K_3} + \cdots}{1 + \dfrac{[A]}{K_1} + \dfrac{[A]^2}{K_1K_2} + \dfrac{[A]^3}{K_1K_2K_3} + \cdots}$$

In order to simplify this quite general expression, we have to derive a relationship between the successive Ks. Now if it is assumed that the sites are independent and equivalent (i.e. that the free energy of interaction of the ligand with each site is the same), then the Ks are related to each other by statistical factors, i.e. A can dissociate from the PA_2 complex in two ways, but A can associate with the $(n-1)$ vacant sites in the PA complex in $(n-1)$ ways. The general relationship is that the ith dissociation constant (K_i) is given by

$$K_i = \left(\frac{i}{n-i+1}\right)K,$$

where K is an intrinsic *dissociation* constant (i.e. one which takes into account these statistical factors).†
Thus

$$K_1 = \frac{K}{n}, \qquad K_2 = \frac{2K}{n-1}, \qquad K_3 = \frac{3K}{n-2} \quad \text{etc.}$$

Our expression for r now becomes

$$r = \cfrac{[A] \cdot \dfrac{(n)}{K} + \dfrac{2[A]^2(n)(n-1)}{2K^2} + \dfrac{3[A]^2(n)(n-1)(n-2)}{(2)(3)K^3} + \cdots}{1 + \dfrac{[A](n)}{K} + \dfrac{[A]^2(n)(n-1)}{(2)K^2} + \dfrac{[A]^3(n)(n-1)(n-2)}{(2)(3)K^3} + \cdots}$$

$$= \cfrac{\dfrac{[A](n)}{K}\left[1 + \dfrac{[A](n-1)}{K} + \dfrac{[A]^2(n-1)(n-2)}{2K^2} + \cdots\right]}{1 + \dfrac{[A](n)}{K} + \dfrac{[A]^2(n)(n-1)}{2K^2} + \dfrac{[A]^3(n)(n-1)(n-2)}{6K^3} + \cdots}.$$

The expressions in the numerator and denominator in this equation are both binomial expansions. Thus the expression can be simplified to give:

$$r = \cfrac{\dfrac{[A](n)}{K}\left(1 + \dfrac{[A]}{K}\right)^{n-1}}{\left(1 + \dfrac{[A]}{K}\right)^n}$$

$$= \cfrac{\dfrac{[A](n)}{K}}{\left(1 + \dfrac{[A]}{K}\right)}.$$

---

† K is actually the geometric mean of all the dissociation constants i.e. $K = (K_1K_2K_3 \ldots K_n)^{1/n}$.

i.e.

$$r = \frac{n[\mathrm{A}]}{K + [\mathrm{A}]}.$$

This is the equation used in Chapter 4 (eqn (4.8)) for the analysis of binding data in the case of multiple ligand binding sites. It is valid provided the sites are equivalent and independent.

Appendix 3

The half-time method for reaction order

CONSIDER the reaction

$$A \rightarrow products,$$

for which the rate law is

$$-\frac{d[A]}{dt} = k[A]^n.$$

Where n is the order. For $n \neq 1$ this can be integrated

$$\frac{[A]^{1-n}}{(1-n)} = kt + c.$$

When $t = 0$, $[A] = [A]_0$

$$\therefore \quad c = \frac{[A]_0^{1-n}}{1-n},$$

$$\therefore \quad kt = \frac{[A]_0^{1-n}}{(1-n)} - \frac{[A]^{1-n}}{(1-n)}.$$

When $t = t_{\frac{1}{2}}$, $[A] = \frac{1}{2}[A]_0$

$$kt_{\frac{1}{2}} = \frac{[A]_0^{1-n}[2^{n-1} - 1]}{(n-1)},$$

$$\therefore \quad t_{\frac{1}{2}} = \frac{[2^{n-1} - 1]}{k(n-1)[A]_0^{n-1}},$$

i.e.

$$t_{\frac{1}{2}} \propto \frac{1}{[A]_0^{n-1}}.$$

When $n = 1$, we have already seen that the expression for the half time is given by

$$t_{\frac{1}{2}} = \frac{\ln 2}{k}. \qquad \text{(see p. 136)}$$

so again

$$t_{\frac{1}{2}} \propto \frac{1}{[A]_0^{n-1}},$$

since there is no dependence on $[A]_0$.

Thus for all values of n,

$$\boxed{t_{\frac{1}{2}} \propto \frac{1}{[A]_0^{n-1}}}.$$

Appendix 4

The interaction of an enzyme with substrate (S) and inhibitor (I)

CONSIDER the scheme below where E, S, and I represent enzyme, substrate, and inhibitor respectively

$$E + S \underset{}{\overset{K_s}{\rightleftharpoons}} ES \overset{k_2}{\rightarrow} E + P$$

$$+ I \qquad\qquad + I$$

$$K_{EI} \uparrow\downarrow \qquad\quad \uparrow\downarrow K_{ESI}$$

$$EI + S \overset{K_s'}{\rightleftharpoons} ESI$$

ESI is assumed to be inactive, and K_s, K_{EI} etc. represent dissociation constants. Now

$$[ES] = \frac{[E][S]}{K_s},$$

and

$$[EI] = \frac{[E][I]}{K_{EI}},$$

and

$$[ESI] = \frac{[ES][I]}{K_{ESI}} = \frac{[E][S][I]}{K_{ESI} \cdot K_s}.$$

The fraction (F) of enzyme in the form of the [ES] complex is given by:

$$F = \frac{[ES]}{[E] + [ES] + [EI] + [ESI]}$$

$$= \frac{\dfrac{[S]}{K_s}}{1 + \dfrac{[S]}{K_s} + \dfrac{[I]}{K_{EI}} + \dfrac{[S][I]}{K_{ESI} \cdot K_s}}.$$

Now the observed velocity (v) is related to the maximum velocity (V_{max}) by:

$$v = V_{max} \cdot F$$

$$= V_{max} \frac{\dfrac{[S]}{K_s}}{1 + \dfrac{[S]}{K_s} + \dfrac{[I]}{K_{EI}} + \dfrac{[S][I]}{K_{ESI} \cdot K_s}}.$$

and this can be rearranged to give:

$$\frac{1}{v} = \frac{1}{V_{max}}\left(1 + \frac{(I)}{K_{ESI}}\right) + \frac{K_s}{V_{max}}\left(1 + \frac{[I]}{K_{EI}}\right) \cdot \frac{1}{[S]}.$$

Solutions to problems

Chapter 1

1. ΔH°_{f} (fumaric) $= -807\cdot6$ **kJ mol^{-1}**
   ΔH°_{f} (maleic) $= -784\cdot4$ **kJ mol^{-1}**
   ΔH°_{f} (maleic $\rightarrow$ fumaric) $= -23\cdot2$ **kJ mol^{-1}.**
   Thus fumaric acid in which the carboxyl groups are *trans* to each other is the more stable isomer.

2. $\Delta H = \Delta U + \Delta n(RT)$.
   For glucose, $\Delta n = 0$, so $\Delta H = -2880$ **kJ mol^{-1}.**
   For stearic acid, $\Delta n = -8$, so $\Delta H = -11\ 381$ **kJ mol^{-1}.**
   Since the molecular weights of glucose and stearic acid are 180 and 284 respectively, this means that the ΔH values are -16 and $-40\cdot1$ kJ g^{-1} respectively. Also fatty acids and fats (esterified fatty acids) require much less associated water for storage, and these two factors make fat a much more efficient energy reserve than carbohydrate. If a bird had to store all its energy in the form of glycogen, it has been calculated that the extra weight would render it incapable of leaving the ground!

3. $\Delta H = -480\cdot7$ **kJ mol^{-1}.**

4. The heat required to warm up the hamster $= (30)(3\cdot3)(100)$ J $= 9\cdot9$ kJ. Now from question 2 combustion of 1 g of fatty acid will provide $40\cdot1$ kJ. So approximately **0·25g** of fatty acid will be required (or 0·25 per cent of the animal's body weight). Measurements of the amount of brown adipose tissue (the fat tissue largely responsible for heat production) suggest that in hibernators it constitutes about 2 per cent of the body weight. This would seem to provide adequate energy reserves for the arousal.

5. (*a*) From the data

$$Mg(s) \;\rightarrow\; Mg^{2+}(g) \qquad \Delta H^{\circ} = +2349\cdot2 \text{ kJ mol}^{-1}$$

and

$$Cl_2(g) \;\rightarrow\; 2Cl^-(g) \qquad \Delta H^{\circ} = -488\cdot2 \text{ kJ mol}^{-1},$$

so that

$$Mg(s) + Cl_2(g) \;\rightarrow\; Mg^{2+}(g) + 2Cl^-(g) \qquad \Delta H^{\circ} = +1861 \text{ kJ mol}^{-1}$$

and using

$$Mg(s) + Cl_2(g) \;\rightarrow\; MgCl_2(s) \qquad \Delta H^{\circ} = -639\cdot5 \text{ kJ mol}^{-1}$$

we have

$$Mg^{2+}(g) + 2Cl^-(g) \;\rightarrow\; MgCl_2(s) \qquad \Delta H^{\circ} = -2500\cdot5 \text{ kJ mol}^{-1},$$

i.e. the lattice energy of MgCl$_2$ is **2500·5 kJ mol^{-1}.**
(*b*) Since MgCl$_2$(s) $\rightarrow$ MgCl$_2$(aq), $\Delta H = -150\cdot5$ kJ mol^{-1}, the heat of hydration of Mg^{2+} + 2Cl$^-$ ions (gaseous) is **-2651 kJ mol^{-1}.**

(c) Assuming that for

$$Cl^-(g) \rightarrow Cl^-(aq) \qquad \Delta H = -383 \cdot 7 \text{ kJ mol}^{-1}$$

we obtain the heat of hydration of $Mg^{2+}(g)$ as $-1883 \cdot 6 \text{ kJ mol}^{-1}$. This value is numerically greater than that for $Ca^{2+}(g)(-1560 \text{ kJ mol}^{-1})$ principally because of the smaller ionic radius, r, of Mg^{2+}. The theory suggests that the $-\Delta H_{\text{hydration}}$ is proportional to r^{-2}. This preferential hydration of Mg^{2+} has important implications in biochemistry. Mg^{2+} usually occurs as a cofactor in enzyme reactions especially those involved in transfer of phosphoryl groups. In these cases the Mg^{2+} ion remains hydrated. However, Ca^{2+} has a very important role in acting as a 'structural cement' in many cases, e.g. in bones, shells, etc. In these cases the Ca^{2+} ion is not hydrated.

6. ΔH for the process $CH_4(g) \rightarrow C(g) + 4H \cdot (g)$ is given by $\Delta H_3^\circ + 2\Delta H_2^\circ - \Delta H_1^\circ$ i.e. $+1663 \cdot 2 \text{ kJ mol}^{-1}$, i.e. the bond energy for C—H is $415 \cdot 8 \text{ kJ mol}^{-1}$. From the data, the ΔH for the process

$$C_2H_4(g) \rightarrow 2C(g) + 4H.(g)$$

can be calculated as $2253 \cdot 1 \text{ kJ mol}^{-1}$.

This process involves breaking four C—H bonds and the C=C bond. Subtracting the contribution of the C—H bonds, this leaves the C=C bond energy as $589 \cdot 9 \text{ kJ mol}^{-1}$.

7. For the hydrogenation of ethene, $\Delta n = -1$, so $\Delta H = -136 \cdot 3 \text{ kJ mol}^{-1}$. Comparing this result with that given for benzene, we calculate that benzene is $200 \cdot 7 \text{ kJ mol}^{-1}$ more stable than would be expected for three isolated double bonds.

Chapter 2

1. (a) $\Delta S = 1 \cdot 22 \text{ J K}^{-1} \text{ g}^{-1}$
      $= 22 \cdot 0 \text{ J K}^{-1} \text{ mol}^{-1}$
   (b) $\Delta S = 6 \cdot 5 \text{ J K}^{-1} \text{ g}^{-1}$
      $= 117 \cdot 0 \text{ J K}^{-1} \text{ mol}^{-1}$

2. The ΔG° values for these reactions are evaluated from the ΔG_f° values.
   (a) $\Delta G^\circ = +3 \text{ kJ mol}^{-1}$, i.e. would not proceed spontaneously.
   (b) $\Delta G^\circ = +13 \text{ kJ mol}^{-1}$, i.e. would not proceed spontaneously.
   (c) $\Delta G^\circ = -27 \text{ kJ mol}^{-1}$, i.e. could proceed spontaneously.
   (d) Clearly this reaction would not proceed spontaneously below $273 \cdot 15 \text{ K}$. For the ice $\rightarrow$ water transition, ΔS is positive (water has a greater degree of randomness than ice). Now at $273 \cdot 15 \text{ K}$, $\Delta H = T \Delta S$ since $\Delta G = 0$ (equilibrium). Thus, assuming ΔH and ΔS are both independent of temperature, at temperatures below $273 \cdot 15 \text{ K}$, ΔG is positive and the transition will not occur spontaneously. When T is above $273 \cdot 15 \text{ K}$, ΔG is negative and ice will melt spontaneously.

3. For

$$ClCH_2CO_2H + OH^- \rightarrow ClCH_2CO_2^- + H_2O; \qquad \Delta H^\circ = -62 \cdot 3 \text{ kJ mol}^{-1}$$

so that for the ionization of $ClCH_2CO_2H$

$$ClCH_2CO_2H \rightarrow ClCH_2CO_2^- + H^+; \qquad \Delta H^\circ = -5 \cdot 5 \text{ kJ mol}^{-1}$$

Since $\Delta G°$ for the ionization is $17 \cdot 1 \text{ kJ mol}^{-1}$ we have $\Delta S°$ ionization $=$ $(\Delta H° - \Delta G°)/T = -75 \cdot 8 \text{ J K}^{-1} \text{ mol}^{-1}$, i.e. the ionization of chloracetic acid is very unfavourable on entropy grounds. The ordering of the solvent molecules by the ions produced (i.e. negative ΔS) is much more significant than the increase in entropy which would be expected from the increase in the number of species on dissociation.

4. From the data, and using the fact that G and H are both state functions, we deduce that for

$$CH_4(\text{inert solvent}) \rightarrow CH_4(\text{aqu})$$

$$\Delta G° = 11 \cdot 7 \text{ kJ mol}^{-1}$$

$$\Delta H° = -11 \cdot 3 \text{ kJ mol}^{-1}$$

From these values we can calculate $\Delta S°$ for the transfer as $-77 \cdot 2 \text{ J K}^{-1} \text{ mol}^{-1}$. This transfer process is thought of as a model for 'hydrophobic' interactions', i.e. to explain why non-polar molecules (or amino acid side chains) prefer non-polar environments to aqueous solution. These values suggest that the 'hydrophobic' effect is primarily due to the entropy term, and this has been explained on the basis that hydrocarbon (or non-polar) molecules would lead to an 'ordering' of water molecules if placed in the aqueous phase (i.e. negative ΔS). In proteins, the amino acids with non-polar side chains are usually 'buried' in the interior of the molecule, away from the solvent water.

5. For the process, $\Delta S°$ is $70 \cdot 2 \text{ J K}^{-1} \text{ mol}^{-1}$. This reaction is therefore dominated by the entropy term, which is positive even though a complex is formed from two species. By reference to the previous example, we might suggest that 'hydrophobic' interactions between the inhibitor and the enzyme are important.

6. $\Delta G°' = -15 \cdot 2 \text{ kJ mol}^{-1}$.

7. For the 'coupled reaction'

$$\text{Creatine phosphate} + \text{ADP} \rightarrow \text{Creatine} + \text{ATP}$$

$$\Delta G°' = -7 \cdot 1 \text{ kJ mol}^{-1}.$$

i.e. the creatine phosphate hydrolysis *could* be used to favour the synthesis of ATP from ADP.

8. $\Delta G° = +43 \cdot 9 \text{ kJ mol}^{-1}$. The values of ΔH and ΔS are very large because they refer to one mole of protein which is a large quantity of protein. Thus for unfolding many hydrogen bonds, etc., must be broken (high $\Delta H°$), but this would result in a great deal of flexibility of the molecule (high $\Delta S°$).

Chapter 3

1. $\Delta G°$ refers to standard state conditions.
(a) $\Delta G° = 0$ At $373 \cdot 15 \text{ K}$ the liquid and gas are in equilibrium, so that $\Delta G = 0$. Since both are in their standard states, $\Delta G°$ also equals 0.

(*b*) $\Delta G° = 7\cdot2$ kJ mol^{-1}. This result can be derived by considering the cycle below

$$H_2O\ (g)$$
$$310\ K$$
(i) $\nearrow$ 0·062 atm $\searrow$ (ii)

$H_2O(l)$ $\qquad\qquad$ $H_2O(g)$
310 K $\qquad\qquad$ 310 K
$\qquad\qquad\qquad$ 1 atm

For process (i) $\Delta G = 0$, since liquid and gas are in equilibrium.
For process (ii)

$$\Delta G = RT\ \ln(P_2/P_1)$$

$$= RT\ \ln\left(\frac{1}{0\cdot062}\right)$$

$$= 7\cdot2\ \textbf{kJ mol}^{-1}.$$

Thus for the overall process linking liquid and gas in their standard states at 310 K we add these two figures. This gives $\Delta G° = \textbf{7·2 kJ mol}^{-1}$. Note that the standard state of the gas at 310 K is not a stable state (but this does not matter).

2. For the reaction $K = \textbf{0·0113 (mol dm}^{-3}\textbf{)}$ from the value of $\Delta G°'$. If x mol dm^{-3} is the concentration of L-glycerol-1-phosphate at equilibrium then

$$\frac{x}{(1-x)(0\cdot5-x)} = 0\cdot0113$$

from which $x = 0\cdot0056$, so that the concentration of L-glycerol-1-phosphate at equilibrium is $\textbf{5·6 mmol dm}^{-3}$.

3. (*a*) For the reaction $K = 0\cdot43$ (mol dm^{-3}), so that at equilibrium

$$[\text{F-6-P}] = \textbf{0·03 mol dm}^{-3}, [\text{G-6-P}] = \textbf{0·07 mol dm}^{-3}.$$

(*b*) For the second reaction, $K = 18\cdot84$ (mol dm^{-3})

$$\frac{[\text{G-6-P}]}{[\text{G-1-P}]} = 18\cdot84, \qquad \frac{[\text{F-6-P}]}{[\text{G-6-P}]} = 0\cdot43$$

and

$$[\text{G-6-P}] + [\text{G-1-P}] + [\text{F-6-P}] = 0\cdot1\ \text{mol dm}^{-3}.$$

The final composition is

$$[\text{F-6-P}] = \textbf{0·029 mol dm}^{-3}, \qquad [\text{G-6-P}] = \textbf{0·0674 mol dm}^{-3}$$

and $[\text{G-1-P}] = \textbf{0·0036 mol dm}^{-3}.$

4. (a) The synthesis of G-6-P can be coupled to the hydrolysis of ATP. For

$$\text{glucose} + \text{ATP} \; \rightleftharpoons \; \text{glucose-6-phosphate} + \text{ADP}$$

$$\Delta G^{\circ\prime} = -13\cdot4 \text{ kJ mol}^{-1},$$

i.e. **$K = 182$ (mol dm^{-3}).**

(b) The synthesis of ATP can be coupled to the hydrolysis of PEP. For

$$\text{PEP} + \text{ADP} \; \rightleftharpoons \; \text{pyruvate} + \text{ATP}$$

$$\Delta G^{\circ\prime} = -24\cdot7 \text{ kJ mol}^{-1},$$

i.e. **$K = 1\cdot46 \times 10^4$ (mol dm^{-3}).**

5. K for the reaction $= 0\cdot31$ (mol dm^{-3}).
So if [Pi] = [G-1-P], [(glycogen)$_{n-1}$]/[(glycogen)$_n$]
$= $ **$0\cdot31$,** i.e. synthesis of glycogen is favoured.
If [Pi] = 10 mmol dm^{-3} and [G-1-P] = 30 μmol dm^{-3},
[(glycogen)$_{n-1}$]/[(glycogen)$_n$] = **102,** i.e. degradation of glycogen is favoured.
   This result shows that in the muscle cell, the levels of Pi and G-1-P are such that phosphorylase acts to degrade glycogen. There is in fact a separate system (involving UDP-glucose as an intermediate) for the synthesis of glycogen. Confirmation of this role of phosphorylase is provided by sufferers from McArdle's disease (where muscle phosphorylase is absent). These patients have high levels of muscle glycogen (which could not be the case if phosphorylase acted in the direction of glycogen synthesis).

6. The ratio

$$\frac{[\text{FDP}][\text{ADP}]}{[\text{F-6-P}][\text{ATP}]} = 0\cdot031$$

and therefore

$$\Delta G' = \Delta G^{\circ\prime} + RT \ln (0\cdot031)$$

$$= -17\cdot7 - 8\cdot89 \text{ kJ mol}^{-1}$$

$$= \mathbf{-26\cdot6 \text{ kJ mol}^{-1}},$$

i.e. the reaction is not at equilibrium (where $\Delta G' = 0$)
The ratio

$$\frac{[\text{AMP}][\text{ATP}]}{[\text{ADP}]^2} = 0\cdot416$$

and therefore

$$\Delta G' = \Delta G^{\circ\prime} + RT \ln (0\cdot416)$$

$$= 2\cdot1 - 2\cdot24 \text{ kJ mol}^{-1}$$

$$= \mathbf{-0\cdot14 \text{ kJ mol}^{-1}},$$

i.e. the reaction is almost at equilibrium.

These results point to the possibility that phosphofructokinase may be a regulatory enzyme in glycolysis. Adenylate kinase is, however, presumably present in sufficient amounts to ensure that equilibrium between the various adenine nucleotides prevails, and is unlikely to represent a control point.

7. $\Delta G°$ for the process below is -13 kJ mol^{-1}

$$\text{Leu}-\text{Gly}+\text{H}_2\text{O} \rightleftharpoons \text{Leu}+\text{Gly}$$

i.e. (amino acid)$_1$ − (amino acid)$_2$ + H$_2$O $\rightleftharpoons$ (amino acid)$_1$ + (amino acid)$_2$ and for the process

$$\text{(amino acid)}_1 + \text{tRNA} \rightleftharpoons \text{(amino acyl)}_1 - \text{tRNA}$$

$\Delta G° = +31·7$ kJ mol^{-1}.
Thus for the chain extension reaction $\Delta G° = -18·7$ kJ mol^{-1}, i.e. **$K = 1·9 \times 10^3$ (mol dm^{-3}).**

This shows that attaching the tRNA to the amino acid has sufficiently 'activated' it to drive the peptide bond formation reaction over to the right.

8. (a) $\Delta G° = -10·8$ kJ mol^{-1}.
   (b) For O$_2$(g) $\rightleftharpoons$ O$_2$(aq)

$$K = \frac{2·3 \times 10^{-4}}{0·2} \text{ (mol dm}^{-3}, \text{atm)}$$
$$= 1·15 \times 10^{-3} \text{ (mol dm}^{-3}, \text{atm)}$$

   $\Delta G° = +16·4$ kJ mol^{-1}.
   (c) Thus for Hb(aq) + O$_2$(aq) $\rightleftharpoons$ HbO$_2$(aq)
   $\Delta G° = -27·2$ kJ mol^{-1}.

The assumptions here are (a) *reversibility* is assumed in the derivation of the equation $\Delta G° = -RT \ln K$; (b) the solutions are assumed *ideal*, i.e. that concentration terms can be used in the equilibrium expression.

9. Using the Van't Hoff isochore we find

   (a) K (298 K) = **$1·78 \times 10^5$ (mol dm^{-3})**
   (b) K (273·15 K) = **$3·71 \times 10^5$ (mol dm^{-3}).**

Note that we assume that $\Delta H°$ and $\Delta S°$ are constant over this temperature range.
   The value of K must be divided by the concentration of water (55 mol dm^{-3}) to obtain the true equilibrium constant for the reaction (K_{eq}). Thus K_{eq} (310 K) = $1·3 \times 10^5/55$ (mol dm^{-3}) = $2·36 \times 10^3$ (mol dm^{-3}). Putting [ADP] = [ATP] and [P$_i$] = 10 mmol dm^{-3}, the concentration of H$_2$O is calculated as **4.2 μmol dm^{-3}.**

10. The plot of $\ln K_d$ vs. $1/T$ is linear. The slope gives $\Delta H° = +20·9$ kJ mol^{-1}. From the graph, when $T = 303$ K, $\ln K_d = -9·99$ (i.e. $K_d = 4·59 \times 10^{-5}$ (mol dm^{-3})) and so $\Delta G°_{303} = +25·2$ kJ mol^{-1}. Hence $\Delta S° = -14·2$ J K^{-1} mol^{-1}. The assumptions are (a) *constancy of* $\Delta H°$ and $\Delta S°$ throughout this temperature range, which seem to be justified by the linearity of the plot; (b) *reversibility* of the process, to use the equality in the second law, in the derivation of the equations.

11. The equilibrium constants for binding are:

	293 K	313 K
Isoleucine tRNA	$2·64 \times 10^7$	$2·64 \times 10^7$
Valine tRNA	$6·89 \times 10^5$	$1·65 \times 10^6$

The factors are thus **38** (293 K) and **16** (313 K).

12. At 293 K, $\Delta G° = -2\cdot8\,\text{kJ mol}^{-1}$, i.e. association to give dimers *would be* favourable.

   On lowering the temperature the equilibrium constant for association would fall (since $\Delta H°$ is positive), i.e. dissociation would be favoured. In many cases, the 'cold lability' of associated enzymes (i.e. inactivation by lowering the temperature) has been shown to arise from dissociation at the low temperatures; the isolated subunits being inactive. Since association is favoured by *entropy* rather than *enthalpy* factors, we could suggest that 'hydrophobic' forces are involved in the association (cf. Chapter 2, problems 4 and 5).

Chapter 4

1. Since ADP is always present in large excess over enzyme, we can treat $[\text{ADP}]_{\text{total}}$ as $[\text{ADP}]_{\text{free}}$. Thus a plot of (no. moles of ADP bound) vs.

$$\left(\frac{\text{no. of moles ADP bound}}{[\text{ADP}]_{\text{total}}}\right)$$

would give a line of slope $(-1/K_d)$ and intercept on the x axis of n. From this we find that $K_d = 6\cdot4\times10^{-4}$ (mol dm^{-3}), $n = 4$. The four sites are equivalent and independent because the binding plot appears linear. (Pyruvate kinase contains 4 subunits, each of which can bind one ADP molecule).

2. (*a*) In the case quoted, both S and I are present in vast excess over E, and their *free* concentrations are effectively equal to their total concentrations. Thus

$$\frac{[\text{E}]}{[\text{ES}]} = \frac{1}{3}, \frac{[\text{E}]}{[\text{EI}]} = \frac{1}{2},$$

   i.e. $[\text{E}] = 0\cdot17[\text{E}]_{\text{total}}$; $[\text{ES}] = 0\cdot5[\text{E}]_{\text{total}}$; $[\text{EI}] = 0\cdot33[\text{E}]_{\text{total}}$.

   (*b*) When $[\text{E}]_{\text{total}}$ becomes comparable with $[\text{S}]_{\text{total}}$ and $[\text{I}]_{\text{total}}$, we can no longer make the assumption above. In this case we have to solve two simultaneous quadratic equations to evaluate the concentrations of all the species. This involves the solution of a cubic equation (usually performed by computer).

3. From plot of fluorescence increase against NADPH added, we can deduce that the enzyme can bind $0\cdot3$ µmol dm^{-3} NADPH. As the enzyme concentration is $0\cdot15$ µmol dm^{-3}, this means there are *two* binding sites for NADPH on the enzyme. From other work the enzyme is known to consist of two subunits. The binding of NADPH is clearly very tight indeed, so that no appreciable dissociation of the enzyme–NADPH complex is observed under these conditions.

4. From either a Scatchard or Hughes–Klotz plot we deduce that there are 4 binding sites (i.e. $n = 4$) and that $K = 1\cdot48\times10^{-4}$ (mol dm^{-3}). The sites appear equivalent and independent. (From other work, the enzyme is known to consist of 4 subunits (i.e. each subunit binds one metal ion)).

5. From a Scatchard or Hughes–Klotz plot we can deduce that the binding shows features typical of positive co-operativity (this should be compared with the binding of NAD$^+$ to the rabbit muscle enzyme which shows features of negative co-operativity). From the extrapolated value of the intercept it is likely that there are four binding sites on the enzyme (which is, in fact, a tetramer).

6. The binding plots show features typical of negative co-operativity. Extrapolation suggests there are six binding sites for CTP per enzyme molecule (i.e. one per regulatory subunit).

Chapter 5

1. Atmospheric pressure = **0·705 atm.**
   The Clapeyron–Clausius equation is used to find the vapour pressure of water at 364·1 K.

2. Molecular weight = **150.**
   If the solute is 50 per cent dissociated we have a 1·5-fold increase in the number of particles present ($i = 1·5$), i.e. the vapour pressure lowering will be 0·000 45 atm (new vapour pressure = 0·024 55 atm).

3. Activity (a) of water (i.e. p_A/p_A^*) = **0·701.**
   Now the mole fraction of water = **0·773**
   so that the activity coefficient = a/X = **0·906.**

4. (1) Freezing point depression = **0·42 K** (note that $(NH_4)_2SO_4$ is completely dissociated to give 3 ions).
   (2) Freezing point depression = **0·000 93 K.**
   (3) Freezing point depression = **0·420 93 K** (i.e. the sum of the contributions from the individual species).
   This result shows that the freezing point depression is too insensitive for the determination of molecular weights of macromolecules.

5. Freezing point depression = **8·3 K.** This result is derived by using the full expression in the derivation of eqn (5.6). If the approximation for dilute solutions is used ($\Delta T_f = 1·86\,(x/M)$) the result is 8·7 K. Both methods assume that the solution is ideal.
   The insects can actually survive exposure to temperatures considerably lower than this, because the haemolymph can be supercooled without freezing.

6. Freezing point depression = **0·12 K** (the concentration of glucose is so low we can equate concentrations expressed in terms of mol kg^{-1} and mol dm^{-3}).

7. (a) $I = $ **0·3**
   (b) $I = $ **0·15**
   (c) $I = $ **0·055**
   (d) $I = $ **0·001 31** (the concentrations of H^+ and acetate$^-$ are both 1·31 mmol dm^{-3})

8. $\log \gamma_\pm = -0·5\, Z_A Z_B \sqrt{(I)}.$
   (a) In 1 mmol dm^{-3} NaCl, $\gamma_\pm = 0·966$, hence $a_{Na^+} = a_{Cl^-} = $ **9·66 × 10^{-4}.**
   (b) In 1 mmol dm^{-3} NaCl and 3 mmol dm^{-3} KHCO$_3$, $\gamma_\pm = 0·93$, hence $a_{Na^+} = a_{Cl^-} = $ **9·3 × 10^{-4}.**

9. When 99 per cent of the Fe^{3+} is precipitated, $[Fe^{3+}] = 0·5 \times 10^{-6}$ mol dm^{-3}. Hence $[OH^-] = 1·26 \times 10^{-10}$, i.e. pH = 4·1.
   Thus **above pH 4·1** more than 99 per cent of the Fe^{3+} would be precipitated. At physiological pH only a very small concentration of Fe^{3+} ($\approx 10^{-15}$ mol dm^{-3}) can remain in solution. This suggests that Fe^{3+} in the plasma must be complexed with other species (in this case, the protein ferritin).

10. At 298 K, the solubility product is **7·84 × 10^{-6}**, hence the solubility of Ca(OH)$_2$ is **0·93 g dm^{-3}.**
    At 275 K, the solubility product is **1·355 × 10^{-5}**, hence the solubility is **1·11 g dm^{-3}.**

11. For this solution $I = 0\cdot01$ (essentially all arising from the added $NaNO_3$). Thus $\gamma_\pm = 0\cdot89$, from the Debye–Hückel Law, and so $[Ag^+] = 1\cdot42 \times 10^{-5}$ g ions dm^{-3}.

In the worked example, there was a common ion (Cl^-) present, and the solubility of AgCl was depressed. This problem illustrates the fact that sparingly soluble salts become more soluble in solutions of high ionic strength, provided no common ion is present.

Chapter 6

1. Molecular weight = **12 600**.

In the presence of a smaller concentration of sodium chloride, the Donnan effect will lead to an unequal distribution of ions across the membrane (unless the protein carries no net charge, i.e. is at its isoelectric point). This would lead to an erroneous molecular weight for the protein. The Donnan effect is negligible if the concentration of salt is much larger than that of the protein.

2. Osmotic pressure difference = **0·027 atm**.

This difference is of great importance in living systems. The balance of this pressure and the hydrostatic pressure in the capillary bed governs the flow of water and small ions from the capillaries to the tissues. A lowering of the protein level in the capillaries (e.g. by starvation) is one of the factors which leads to an accumulation of water in the tissues (a condition known as hypoproteinoemic oedema).

3. The total concentration of dissolved species is **0·3 mol dm^{-3}**. The freezing point depression of the plasma is **0·56 K**. If all small solutes were removed the freezing point depression exerted by the protein would be **0·002 K**.

Now 0·95 per cent NaCl is 0·16 mol dm^{-3}, but since NaCl is dissociated to give two ions, the total concentration of dissolved species is **0·32 mol dm^{-3}**, i.e. the osmotic pressure of this solution is the same as that exerted by the plasma.

4. Osmotic pressure difference = **0·134 atm**.

This large osmotic pressure difference has pronounced effects in the diabetic patient. The high osmotic pressure of extracellular fluids necessitates a large urine volume, and would lead to loss of water from the tissues. A severely diabetic patient remains thirsty in spite of drinking large amounts of water and would suffer severe dehydration if he were untreated.

5. Freezing point depression should be **0·0011 K**.

The freezing point depression is some 550-fold greater than this and indicates the effectiveness of the 'anti-freeze protein'. The exact mechanism of this remains unsolved (see text).

The calculated freezing point depression for lysozyme (0·0013 K) is in close agreement with the observed value and indicates that the solution is behaving almost ideally.

6. The equilibrium concentrations are: inside $[Na^+] =$ **79·5 mmol dm^{-3}**; $[Cl^-] =$ **73·5 mmol dm^{-3}**; outside $[Na^+] = [Cl^-] =$ **76·5 mmol dm^{-3}**. This result shows that the Donnan effect leads to a small difference between the ion concentrations on the two sides of the membrane. In the kidney, however, there is an accumulation of Na^+ and Cl^- on the plasma side of the membrane. This 'active transport' requires the expenditure of energy (ATP hydrolysis), and ensures the minimum loss of ions by excretion.

7. Using the relation $|\Delta G| = RT \ln([c]_{\text{inside}}/[c]_{\text{outside}})$ for each ion the free energy requirements are **6·6 kJ g ion^{-1}** (Na$^+$) and **8·1 kJ g ion^{-1}** (K$^+$). The accumulation of K$^+$ at the expense of Na$^+$ is of crucial importance in living systems, e.g. in the conduction of nerve impulses.

8. (a) Calculate the values of $\ln r$. A plot of $\ln r$ vs. t gives a straight line with slope $= 1\cdot3374 \times 10^{-5}$ s^{-1}. Noting that $\omega^2 = 2\cdot7410 \times 10^7$ s^{-2} then from eqn (6.8)

$$s_{298,\text{w}} = 5\cdot027 \times 10^{-13} \text{ s.}$$

The correction for viscosity has to be made

$$s_{293,\text{w}} = (\eta^\circ_{298}/\eta^\circ_{293}) \cdot s_{298,\text{w}}$$

from which

$$s_{293,\text{w}} = 4\cdot46 \times 10^{-13} \text{ s.}$$

(b) From the Svedberg equation $M = 52\ 400$.

9. (a) From eqn (6.11) noting that $d(\ln c) = \ln c_b - \ln c_m$ and $dr^2 = r_b^2 - r_m^2$ then for $c_b/c_m = 4$, r.p.m. $= 2\cdot45 \times 10^6/\sqrt{M}$.
   If $M = 100\ 000$ then r.p.m. $= 7\cdot74 \times 10^3$
   (b) If $c_b/c_m = 1000$, r.p.m. $= 5\cdot49 \times 10^6/\sqrt{M}$. If $M = 100\ 000$ then r.p.m. $= 1\cdot73 \times 10^4$
   (c) Calculate the r.p.m. values for molecular weights of 10 000 and 1 000 000 as typical

M	High speed (r.p.m.)	Low speed (r.p.m.)
10 000	$5\cdot49 \times 10^4$	$2\cdot45 \times 10^4$
1 000 000	$5\cdot49 \times 10^3$	$2\cdot45 \times 10^3$

Use then the results from (a) and (b) to give the required plot. Note its linearity. This plot is useful for selection of rotor speeds in general.

10. (a) Plot $\ln \Delta I_i$ versus r_i^2 and obtain the slope $= 2\cdot83$.
    Note that

$$\frac{d \ln \Delta I_i}{dr^2} \equiv \frac{d \ln c}{dr^2}.$$

Using eqn (6.11) $M = 128\ 000$
(b) The upwards curvature in the $\ln \Delta I_i$ versus r_i^2 plot near to the bottom of the cell indicates that some high molecular weight contamination and aggregates are present in the sample.

11. (a) Macromolecules—independent of their molecular weight—will float in solvents of greater density than theirs. The density of the lipoprotein component is less than the solvent here and hence migrates upwards.
    (b) Vary the solvent density. This can be done by adding salt, sucrose or D_2O to the lipoprotein solution. The sedimentation rate will not be proportional to the molecular weight for components of different densities and so, by

adjusting the density of the solution, one component will sediment and the other float giving good separation.

12. From the sedimentation velocity experiment alone it is impossible to decide whether unfolding or dissociation or both occur. Additional data such as viscosity, diffusion coefficient etc are necessary to make any firm statements.

    The best way to solve the problem is to determine the molecular weight in the presence and absence of SDS by sedimentation equilibrium experiments.

13. By plotting the data in an appropriate form (elution volume vs. log(molecular weight)) it can be deduced that the molecular weights of creatine kinase and glyceraldehyde-3-phosphate dehydrogenase are 81 000 and 138 000 respectively. Note that the graph curves off at both ends of the molecular weight range (where complete inclusion or complete exclusion of the proteins would be expected).

    The *subunit* molecular weight can be deduced from the electrophoresis experiment (SDS dissociates multi-subunit proteins). From a plot of log(molecular weight) against mobility we can deduce that the subunit molecular weights of creatine kinase and glyceraldehyde-3-phosphate dehydrogenase are 41 000 and 37 000 respectively. From these results we would suggest that the enzymes consist of *two subunits* and *four subunits* respectively. These subunits are probably identical in each case but this would require further data (e.g. on the amino acid sequence) for confirmation.

Chapter 7

1. $K_w(310 \text{ K}) = 2 \cdot 57 \times 10^{-14}$
    This is a straightforward application of the van't Hoff isochore, noting that $\Delta H°$ for the equilibrium $H_2O \rightleftharpoons H^+ + OH^-$ is $+56 \cdot 8 \text{ kJ mol}^{-1}$. Neutral pH at 310 K is $6 \cdot 79$ (i.e. when $H^+ = OH^-$)

2. (*i*) pH = $1 \cdot 3$
    (*ii*) pH = $2 \cdot 88$ $([H^+] = [\text{acetate}^-] = 1 \cdot 31 \times 10^{-3} \text{ mol dm}^{-3})$
    (*iii*) pH = $8 \cdot 79$ $([OH^-] = [C_6H_5\overset{+}{N}H_3] = 6 \cdot 18 \times 10^{-6} \text{ mol dm}^{-3})$

    (*iv*) pH = $2 \cdot 72$
    We derive this result by noting that HCl is completely dissociated and thus contributes $10^{-3} \text{ mol dm}^{-3}$ H^+ to the solution. This partially suppresses the ionization of the acetic acid. If the concentration of acetic acid dissociated is $x \text{ mol dm}^{-3}$, then $[\text{acetate}^-] = x \text{ mol dm}^{-3}$, $[\text{acetic acid}] = (0 \cdot 1 - x) \text{ mol dm}^{-3}$ and $[H^+] = (10^{-3} + x) \text{ mol dm}^{-3}$. Inserting these values in the dissociation constant expression, we obtain x and hence the total $[H^+]$.
    (*v*) pH = $6 \cdot 98$.
    This result is derived in a similar manner to part (*iv*) above. The HCl contributes $10^{-8} \text{ mol dm}^{-3}$ H^+. However the ionization of the water (to give H^+ and OH^-) is suppressed. We add the H^+ contributions from the HCl and the H_2O to obtain the final pH.

3. $\gamma_{H^+} = 0 \cdot 83$ $(a_{H^+} = 8 \cdot 31 \times 10^{-3}, [H^+] = 10^{-2})$.

4. The pH at which the overall charge is zero.
    (*a*) pI = $5 \cdot 59$
    (*b*) pI = $3 \cdot 32$
    (*c*) pI = $9 \cdot 8$
    (*d*) pI = $5 \cdot 02$.

Note that in each case the pI is the average of the two pK_as of the *zwitterion* form.

5. $pH = pK + \log\dfrac{[\text{base}]}{[\text{acid}]}$.

From this the ratio $[H_2PO_4^-]/[HPO_4^{2-}] = 1\cdot585$ and thus **61·3** cm^3 of the **NaH$_2$PO$_4$** and **38·7** cm^3 of the **Na$_2$HPO$_4$** solution are required.

For $0\cdot1$ mol dm^{-3} NaH$_2$PO$_4$, $I = \mathbf{0\cdot1}$.
For $0\cdot1$ mol dm^{-3} Na$_2$HPO$_4$, $I = \mathbf{0\cdot3}$.
For the final buffer, $I = \mathbf{0\cdot177}$.

6. A buffer solution is a solution of a weak acid and its conjugate base which can resist changes in pH on addition of H$^+$ or OH$^-$ ions.

The concentration of sodium acetate to be added is **0·044** mol dm^{-3}.

(i) The pK_a of acetic acid will change with temperature (in accordance with the van't Hoff isochore).

(ii) Increasing the ionic strength will increase the dissociation of the acetic acid (i.e. lower the pK_a). This arises from the decrease in activity coefficients at high ionic strength (Chapter 5).

7. The solution acts as the most effective buffer when pH = pK_a (at this pH, $\alpha = 0\cdot5$ and the expression $\alpha(1-\alpha)$ is a maximum).

For $H_2PO_4^-/HPO_4^{2-}$, pH = **7·2** = pK_a.
For tris H$^+$/tris, pH = **8·08** = pK_a.

(a) **0·499**.
(b) **0·192**.
(c) **0·1**.

The results in parts (a) and (b) are derived by evaluating α (and hence $\alpha(1-\alpha)$) at the various pH values, using the Henderson–Hasselbalch equation. In part (c), α is the same, but C (the concentration) of buffer has decreased.

The results clearly show that a buffer should only be used within a small range (normally ±1 unit) of pH around the pK value, and should be as concentrated as possible.

8. $0\cdot1$ mol dm^{-3} buffer, new pH = **7·98**.
$0\cdot01$ mol dm^{-3} buffer, new pH = **7·82**.

The Henderson–Hasselbalch equation is used to derive the concentrations of tris H$^+$ and tris at pH $8\cdot0$. Addition of 1 mmol dm^{-3} H$^+$ from the enzyme reaction will increase [tris H$^+$] by 1 mmol dm^{-3} and decrease [tris] by this amount. The new pH can then be calculated.

If there were no buffer present, we might expect the final [H$^+$] to be 1 mmol dm^{-3} (i.e. pH = 3). However of course the ADP and P$_i$ present will act as buffers and we could calculate the new pH approximately by assuming that these were equivalent buffers with a pK_a of $7\cdot2$ (see Question 5). If the total $[P_i] = 2$ mmol dm^{-3} the final pH after the reaction would be **6·96**.

9. $[HCO_3^-] = \mathbf{24\cdot00}$ mmol dm^{-3}
$[CO_2] = \mathbf{0\cdot95}$ mmol dm^{-3}

On acidification, the total CO$_2$ (i.e. CO$_2$ + HCO$_3^-$) is measured. This is $24\cdot95$ mmol dm^{-3}. The ratio $[HCO_3^-]/[CO_2]$ is given by the Henderson–Hasselbalch equation.

$$[P_{CO_2}] = \mathbf{0\cdot0306}\ \text{atm}.$$

10. Plotting the titration data we see the pH is changing most rapidly in the pH range 7–10 (i.e. at the equivalence point). The most suitable indicators would thus be *neutral red, phenolphthalein*, or *thymolphthalein*.

11. The plots of log (mobility) vs. log (molecular weight) for the two charge classes (i.e. singly charged and doubly charged) are both linear with a slope of $-\frac{2}{3}$, in accordance with the equation.

The peptides might be expected to be of general formula $(Asp\text{-}Leu)_n$. These would have a charge of n and a molecular weight of approximately $240n$. From the plots, we see that the peptide of mobility $0\cdot75$ (i.e. $\log m = -0\cdot125$) would have a molecular weight 480 if its charge were 2 and a molecular weight of 190 if its charge were 1. Clearly the peptide is $(\mathbf{Asp\text{-}Leu})_2$.

The peptide of mobility $0\cdot45$ (i.e. $\log m = -0\cdot35$) is more of a problem. A peptide of this mobility with a charge of 2 would have a very high molecular weight (approximately 1500), and with a charge of 1 would have a molecular weight of 480. Bearing in mind that asparagine (with no net charge) is converted to aspartic acid (with a charge of 1) on acid hydrolysis, it could be proposed that the probable formula of the peptide is $\mathbf{Asp\text{-}Asn\text{-}Leu_2}$ (Asn $\equiv$ asparagine).

This problem illustrates some of the difficulties involved in deciding whether a certain amino acid residue in a sequence is either aspartic acid or asparagine. A similar problem occurs with glutamic acid and its amide, glutamine.

12. Using the van't Hoff isochore, we find that
For tris

$$\frac{K_a(273)}{K_a(298)} = \mathbf{0\cdot182}$$

This means that the change in pK_a on going from 298 K to 273 K is $+\mathbf{0\cdot74}$ units.
For the phosphate dianion

$$\frac{K_a(273)}{K_a(298)} = \mathbf{0\cdot865},$$

i.e. the change in pK_a is $+\mathbf{0\cdot067}$ units on going from 298 K to 273 K.

This shows that the pH of tris buffers is far more sensitive to changes in temperature than is the pH of phosphate buffers, and this should be borne in mind when using buffers at a variety of temperatures.

Chapter 8

1. (a) $Zn^{2+} + 2e^- \rightarrow Zn.$
   (b) $H^+ + e^- \rightarrow \frac{1}{2}H_2.$
   (c) $Co^{3+} + e^- \rightarrow Co^{2+}.$
   (d) $AgBr + e^- \rightarrow Ag + Br^-.$
   (e) $\frac{1}{2}Hg_2Cl_2 + e^- \rightarrow Hg + Cl^-.$
   Note that in (d) and (e) we can obtain the reaction by adding two reactions, e.g.
   $AgBr(s) \rightarrow Ag^+ + Br^-$ and $Ag^+ + e^- \rightarrow Ag(s).$

   (f) $\begin{matrix} CHCO_2^- \\ \| \\ CHCO_2^- \end{matrix} + 2H^+ + 2e^- \rightarrow \begin{matrix} CH_2CO_2^- \\ | \\ CH_2CO_2^- \end{matrix}$

   fumarate   succinate

Note that H^+ acts as one of the reactants (i.e. reduction by H_2 can be thought of as reduction by $2H^+ + 2e^-$).

(g) Cyt c $(Fe^{3+}) + e^- \rightarrow$ Cyt c (Fe^{2+}).
(h) $CO_2 + H^+ + 2e^- \rightarrow HCO_2^-$.
(i) $NAD^+ + H^+ + 2e^- \rightarrow NADH$.
This is equivalent to reduction of NAD^+ by the hydride ion $(H^- = H^+ + 2e^-)$.

2. The electrochemical cell reaction is given by

Left (reduced) + Right (oxidized) $\rightarrow$ Left (oxidized) + Right (reduced)

and $E°$ by $E°$ (right) $- E°$ (left).

(a) $Cu + Zn^{2+} \rightarrow Cu^{2+} + Zn$, $E° = -1·1$ V.
(b) $\frac{1}{2}H_2 + Ag^+ \rightarrow H^+ + Ag$, $E° = 0·8$ V.
(c) $\frac{1}{2}H_2 + AgCl \rightarrow H^+ + Cl^- + Ag$, $E° = 0·22$ V.
(d) $\frac{1}{2}H_2 + Fe^{3+} \rightarrow H^+ + Fe^{2+}$, $E° = 0·77$ V.
(e) $NADH + oxaloacetate^{2-} + 2H^+ \rightarrow NAD^+ + H^+ + malate^{2-}$, $E°' = 0·15$ V.
We have written the electrochemical cell this way to emphasize that $2H^+$ are involved in the oxaloacetate^{2-}, malate^{2-} half cell and one H^+ in the NAD^+, NADH half cell. *Note* that $2e^-$ are involved in both half cells (see the answer to Question 1).

3. The electrochemical cell reaction is given by

Left (reduced) + Right (oxidized) $\rightarrow$ Left (oxidized) + Right (reduced).

(a) The left-hand half cell is $Pb^{2+}|Pb$ and the electrochemical cell is thus:

$$Pb^{2+}|Pb\|Sn^{2+}|Sn.$$

(b) $Pt|pyruvate^- + 2H^+, lactate^-\|NAD^+ + H^+, NADH|Pt$.

See the comment on Question 2, part (e), regarding the involvement of H^+ ions in this reaction.

4. The electrochemical cell is said to be *reversible* since the reaction can be made to proceed in either direction (see Chapters 2 and 8 for further details).

(i) $$\Delta G° = \Delta H° - T \Delta S°$$
$$= -151·4 \text{ kJ mol}^{-1}$$

Hence the reaction would *proceed in the direction as written* (i.e. left $\rightarrow$ right).
(ii) Since $\Delta G° = -nFE°$ and $n = 2$,

$$E° = 0·78 \text{ V}.$$

Thus an e.m.f. of this magnitude must be applied to prevent this reaction occurring (when the components are in their standard states).
(iii) If the applied e.m.f. is *greater* than 0·78 V the reaction proceeds from *right to left*, and vice versa.

5. The cell reaction is

$$Zn + 2 Fe^{3+} \rightarrow Zn^{2+} + 2 Fe^{2+}.$$

Since $\Delta G^\circ = -nFE^\circ$ and $n = 2$

$$\Delta G_{298}^\circ = -295 \cdot 3 \text{ kJ mol}^{-1}$$

Now

$$\Delta S^\circ = -\frac{d(\Delta G^\circ)}{dT}$$

$$= nF\frac{d(E^\circ)}{dT}$$

$$= 154 \cdot 4 \text{ J K}^{-1} \text{ mol}^{-1}$$

$$\therefore \quad \Delta H^\circ = -249 \cdot 3 \text{ kJ mol}^{-1}$$

The assumptions made are that ΔH° and ΔS° are *both independent of temperature* over this range.

6. For a half cell, the Nernst equation becomes

$$E = E^\circ + \frac{RT}{nF}\ln\frac{[\text{Oxidised}]}{[\text{Reduced}]} \qquad \text{(see text)}$$

or at pH 7

$$E' = E^{\circ\prime} + \frac{RT}{nF}\ln\frac{[\text{Oxidized}]}{[\text{Reduced}]}. \qquad (n = 1 \text{ in this case})$$

E					
(Oxidized/Reduced)	0·3	0·25	0·2	0·15	0·1
	33·4	4·75	0·68	0·097	0·014

Because of the logarithmic nature of the Nernst equation, even quite large errors in the ratio [oxidized]/[reduced] lead to relatively small changes in E (compare the relationship between ΔG° and K, Chapter 3).

7. E° (at pH 0) is related to $E^{\circ\prime}$ (at pH 7) by the equation

$$E^{\circ\prime} = E^\circ + \frac{RT}{2F}\ln\frac{[\text{NAD}^+][\text{H}^+]}{[\text{NADH}]}.$$

(*i*) Thus at pH 0, $E^\circ = -0 \cdot 11$ V.
     Using this value of E° (pH 0) we can then compute the e.m.f. at pH 6.

$$\text{e.m.f. (pH 6)} = -0 \cdot 29 \text{ V}.$$

(*ii*) A similar approach gives E° (pH 0) for this half cell as $0 \cdot 239$ V. Note that there are *two* protons involved here, so the Nernst Equation has a term in $[\text{H}^+]$. Thus the e.m.f. at pH 6 is $-0 \cdot 116$ V.

(*iii*) The reaction is

$$\text{NAD}^+ + \text{H}^+ + \text{malate}^{2-} \rightarrow \text{NADH} + \text{oxaloacetate}^{2-} + 2\text{H}^+.$$

At pH 7, $E^{\circ\prime}$ for this reaction is $-0 \cdot 145$ V.

Now $\ln K' = -\Delta G^{\circ\prime}/RT = nFE^{\circ\prime}/RT$, hence

$$K'(\text{pH 7}) = 1 \cdot 24 \times 10^{-5} \text{ (mol dm}^{-3}, \text{ pH 7)}$$

At pH 6, the e.m.f. is $-0 \cdot 174$ V hence $\ln K$ (pH 6) $= nFE/RT$

$$K \text{ (pH 6)} = 1\cdot29 \times 10^{-6} \text{ (mol dm}^{-3}, \text{pH 6)}$$

Thus by increasing $[H^+]$ (i.e. lowering the pH from 7 to 6) the equilibrium is driven towards the left. The K referred to here (at any given pH) does not include the term in $[H^+]$ since this is already taken into account in converting E° (pH 0) to E at any other pH.

$$\text{i.e. this } K = \frac{[\text{NADH}][\text{oxaloacetate}^{2-}]}{[\text{NAD}^+][\text{malate}^{2-}]}.$$

If we included the term in $[H^+]$ in the expression for K, K would be independent of pH. Thus the true equilibrium constant in this case would be $1\cdot25 \times 10^{-12}$ (mol dm^{-3}).

8. The cell reaction is

$$\tfrac{1}{2}H_2 + \tfrac{1}{2}HgCl_2 \rightleftharpoons Hg + H^+ + Cl^-.$$

From the Nernst equation (since $n = 1$),

$$E = E^\circ - \frac{RT}{F} \ln \frac{a_{H^+} a_{Cl^-}}{(p_{H_2})^{\frac{1}{2}}}$$

If we assume the solution to be ideal and replace activity by concentration for H^+ and Cl^-, we obtain (since $p_{H_2} = 1$):

$$c_{H^+} \cdot c_{Cl^-} = 8\cdot68 \times 10^{-5}$$

$$c_{H^+} = 9\cdot31 \times 10^{-3} \text{ (mol dm}^{-3}) \ (= c_{Cl^-}).$$

Thus the pH of the solution is **2·03**.

9. The coupled reaction

$$\text{Cyt f (Fe}^{3+}) + \text{Cyt b (Fe}^{2+}) \rightleftharpoons \text{Cyt f (Fe}^{2+}) + \text{Cyt b (Fe}^{3+})$$

has an $E^{\circ\prime}$ of $0\cdot3$ V.

For a *one* electron transfer (i.e. one mole of each component involved), $\Delta G^{\circ\prime} = -29$ kJ mol^{-1}.

This *would not be sufficient* to drive the synthesis of ATP from ADP and P_i.

However, if *two* electrons were pased along the redox chain (i.e. two moles of each component involved), $\Delta G^{\circ\prime}$ would be $-57\cdot9$ kJ and this would clearly *be sufficient* to drive the synthesis of *one* mole of ATP from ADP and P_i.

10. We can use eqn (8.3) to calculate $\Delta G' = \mathbf{57\cdot2\ kJ\ mol^{-1}}$. From eqn (8.4) $\Delta p = -592/n$ mV.

We can also estimate Δp from eqn (8.2) if we know $\Delta\psi$ and ΔpH. The calculation of ΔpH is analogous to the worked example on p. 126 and $\mathbf{\Delta pH = +1\cdot66}$ with the inside being more alkaline. $\Delta\psi$ can be obtained from the ^{86}Rb$^+$ distribution data using eqn (8.6) and $\mathbf{\Delta\psi = -118\ mV}$.

From eqn (8.2) $\mathbf{\Delta p = -216\ mV}$.

Comparison with the first value of Δp gives $\mathbf{n = 2.7}$. This means that either the chemiosmotic theory is not an adequate description or that the stoichiometry of proton movement is actually higher than the expected value of $2\cdot0$.

These data represent typical experimental results and it is not difficult to see why the mechanism of oxidative phosphorylation remains such a controversial topic.

Chapter 9

1. *First order.* $k = \mathbf{0\cdot01205}$ min^{-1}.
   The total CO_2 evolved from this solution would be $2\cdot45$ cm^3. From the successive half times, the reaction can be seen to be first order. k can be determined from the half time or by plotting

$$\ln\left(\frac{[P]_\infty}{[P]_\infty - [P]_t}\right) \text{ vs. } t.$$ (see p. 136)

($[P]_t$ is the amount of CO_2 evolved at time t.)

2. Half life of $A = \mathbf{3\cdot3}$ min.
   $B = 0\cdot05$ mol dm^{-3} after $t = \mathbf{5\cdot73}$ min.
   Maximum concentration of B is $\mathbf{0\cdot07}$ mol dm^{-3}.
   These results are derived by noting that $(-d[A]/dt) = (k_1 + k_2)[A]$ (where k_1 and k_2 are the rate constants for the formation of B and C respectively), and that $d[B]/dt = k_1[A]$. Since $[A]_t = [A]_0\, e^{-(k_1+k_2)t}$ we can substitute in the second equation to obtain:

$$[B]_t = \frac{k_1[A]_0}{k_1 + k_2}[1 - \exp\{-(k_1 + k_2)t\}].$$

In this example $k_1 = 0\cdot15$ min^{-1}, $k_2 = 0\cdot06$ min^{-1}.
The maximum concentration of B is found by evaluating $[B]$ at $t = \infty$.

3. The successive half times for the reaction are 27 s, 54 s and 120 s, and the reaction is thus *second order*. We can evaluate the rate constant from this directly ($k = \mathbf{37}$ (mol dm^{-3})$^{-1}$ s^{-1}) or by making an appropriate linear plot.
   For a second order reaction where the concentrations of the components are equal this is

$$\left(\frac{([A]_0 - [A]_t)}{([A]_0)([A]_t)}\right) \text{ against } t.$$ (see p. 138)

($[A]_t$ is the concentration of A remaining at time t, i.e. $([A]_0 - [A]_t)$ is the amount of product formed.) The slope of this line gives k directly.

4. Since the reaction is second order (from Question 3) we plot the appropriate second order expression against time (eqn (9.2)). From this the value of k is found to be $\mathbf{30}$ (mol dm^{-3})$^{-1}$ s^{-1}. Note that we obtain the amount of product formed at time t by subtraction of the N-acetylcysteine concentration at time t from its initial value.

5. By inspection of the decay curve the reaction is seen to be *first order* in A (the successive half times are 39 s, 41 s, and 37 s). Thus $k = \mathbf{0\cdot018}$ s^{-1}.
   This reaction is carried out under pseudo-first-order conditions ($[B] \gg [A]$). If the initial concentration of B is halved the pseudo first order rate constant is halved (i.e. $k = 0\cdot009$ s^{-1}) and the new half life of A is $\mathbf{78}$ s.
   The second-order rate constant, k_2, is obtained by dividing the pseudo first order rate constant by the concentration of the compound in excess. Thus $k_2 = \mathbf{0\cdot018}$ (mol dm^{-3})$^{-1}$ s^{-1} at 300 K. The rate constant at 323 K is found by using the Arrhenius equation, k_2 at 323 K $= 0\cdot193$ (mol dm^{-3})$^{-1}$ s^{-1}.
   Thus when the concentrations of A and B are both $0\cdot1$ mol dm^{-3} the rate of the reaction is $\mathbf{0\cdot00193}$ mol dm^{-3} s^{-1}. (Rate $= k_2[A][B]$.)

6.

$$\frac{d[P]}{dt} = \frac{k_2[S]}{\{(k_{-1}+k_2)/k_1\}+[S]} ([E]_{total}).$$

The outline of the derivation of this equation is given in Chapter 10.

For the steady state approximation to hold, the rate of breakdown of ES to give P must be comparable with or slower than the breakdown to regenerate $E+S$ and $[E]$ must be much smaller than $[S]$. This will then keep $[ES]$ small and reasonably constant, as required by the approximation. The maximum rate of production of P will occur when $[S]$ is large compared with the term $\{(k_{-1}+k_2)/k_1\}$, i.e.

$$\left(\frac{d[P]}{dt}\right)_{max} = k_2[E]_{total}.$$

7. $E_a = 69\cdot3\ \text{kJ mol}^{-1}$
$k_{283} = 3\cdot43\ (\text{mol dm}^{-3})^{-1}\ \text{min}^{-1}.$

8. From the graph of $\ln k$ vs. $1/T$

$E_a = 30\cdot9\ \text{kJ mol}^{-1}.$

9. The plot of $\log_{10}(k/k_0)$ vs. $\sqrt{I}$ is a straight line of slope -3. Since $Z_A \cdot Z_B = -3$, *the data are consistent with the equation.* In the reverse reaction, one of the species is uncharged, and there is *no dependence of the rate constant* on the ionic strength.

10. A plot of $\log k_{rel}$ vs. pH is linear with a slope of $+1$ up to pH $\approx 8\cdot5$. This indicates that the unprotonated form of the amino acid (i.e. $NH_2CHRCO_2^-$) is the reactive species in the nucleophilic displacement reaction. Above pH 10, the k_{rel} begins to level off. The pK_a of the amino group is determined by the point of intersection of the linear portions, i.e. $pK_a = 9\cdot5$.

11. For each buffer we can calculate the concentrations of H^+, acetate$^-$, and acetic acid (see Chapter 7). From this it can be seen that the rate constant does not depend on the concentrations of either H^+ or acetic acid, but is linearly dependent on the acetate$^-$ concentration according to the equation; rate constant $= 0\cdot38 \times 10^{-3} + 0\cdot5$ [acetate$^-$]. This reaction is therefore an example of general base catalysis. (Catalysis by OH^- can be detected at higher pH values).

12. $\Delta S^{o\ddagger} = 58\cdot6\ \text{J K}^{-1}\ \text{mol}^{-1}$. This indicates that the transition state is less ordered than the reactant and this situation is typical for unimolecular decompositions.

Chapter 10

1. *Competitive inhibition.* K_m is raised, but V_{max} is unchanged, so the line will have the same slope but a point of intersection on the x-axis further to the left of the line in the absence of inhibitor.
   Non-competitive inhibition. K_m is unchanged, but V_{max} is decreased, so the line will have the same point of intersection on the x-axis but an increased slope.
   Uncompetitive inhibition. K_m is increased and V_{max} is decreased, so the line will intersect the x-axis further to the left and have an increased slope.

2. Applying the standard eqn (10.3) we see that for the *fasting rat*, v (hexokinase) $= 0\cdot691\ \mu\text{mol min}^{-1}\ \text{g}^{-1}$ and v (glucokinase) $= 0\cdot992\ \mu\text{mol min}^{-1}\ \text{g}^{-1}$, i.e. the glucokinase contribution is **59 per cent** of the total.

For the *well-fed rat*, v (hexokinase) $= 0\cdot697\ \mu\text{mol min}^{-1}\,\text{g}^{-1}$ and v (*glucokinase*) $= 2\cdot095\ \mu\text{mol min}^{-1}\,\text{g}^{-1}$, i.e. the glucokinase contribution is **75 per cent** of the total.

Thus glucokinase can alter its activity in response to changes in the levels of glucose, whereas hexokinase is effectively working at its maximum rate under these conditions. After a carbohydrate intake (e.g. a meal) the glucokinase in the liver can be used to phosphorylate a large proportion of the extra glucose.

3. From an appropriate plot we deduce that in the absence of AMP the K_m for FDP is **8 μmol dm^{-3}** and the V_{max} is **0·185 katal kg^{-1}**. AMP behaves as a *simple non-competitive inhibitor*, with V_{max} lowered to **0·102 katal kg^{-1}**. From eqn (10.6) we can calculate that K_{EI} is **9·8 μmol dm^{-3}**.

4. From a primary plot ($1/v$ vs. $1/[\text{dGDP}]$) we can deduce that the reaction apparently follows a 'ping-pong' mechanism. The kinetic parameters in eqn (10.9) are $K_{dGDP} = $ **49 μmol dm^{-3}**, $K_{GTP} = $ **107 μmol dm^{-3}**, and $V_{max} = $ **0·79 katal kg^{-1}**.

A likely mechanism would be reaction of the enzyme with GTP to form a phosphorylenzyme and GDP. The phosphorylenzyme would then react with dGDP to give dGTP. Tests which could be done include isolation of the presumed phosphorylenzyme and examination of partial reactions.

5. From the primary and secondary plots we deduce values of $K_{ATP} = $ **0·28 mmol dm^{-3}**, $K'_{ATP} = $ **1·7 mmol dm^{-3}**, $K_{creatine} = $ **8·5 mmol dm^{-3}**, and $V_{max} = $ **3·03 katal kg^{-1}**. Referring to the scheme for the random ternary complex mechanism this means that the binding of ATP (strictly speaking MgATP) to the enzyme is enhanced by the prior binding of creatine and vice versa. This is an example of *substrate synergism*.

6. From the appropriate plots we deduce that pyruvate acts as a competitive inhibitor towards lactate and a non-competitive inhibitor towards NAD$^+$. Data such as this has been used to indicate that this enzyme obeys an ordered ternary complex mechanism with NAD$^+$ binding first.

7. The K_m for PEP is **0·4 mmol dm^{-3}** and V_{max} is **2·95 katal kg^{-1}**. 2-Phosphoglycerate acts as a *competitive inhibitor* towards PEP ($K_{EI} = $ **11·5 mmol dm^{-3}**). The close structural similarity between these two molecules makes it likely that they bind at the same site on the enzyme.

Phenylalanine inhibition is more complex and does not belong to any of the three simple categories mentioned. It might be suspected that phenylalanine binds to a site distinct from the substrate site, since there is little structural similarity between the molecules. Present evidence supports this idea and suggests that the inhibition is mediated via changes in the three dimensional structure of the enzyme.

8. From the Arrhenius plot it is seen that there are three distinct phases. (i) $T < 293$ K—the activation energy is **220 kJ mol^{-1}** (ii) $315 > T > 293$ K—the activation energy is **65 kJ mol^{-1}**; (iii) $T > 315$ K the enzyme loses activity, presumably because of thermal denaturation. The transition at 293 K may well be due (in this case) to changes in the phospholipids associated with this enzyme.

9. From a plot of log (V_{max}) vs. pH it is seen that the pK_a (298 K) of the ionizing group is **6·25**. By applying the van't Hoff isochore the enthalpy of ionization can be calculated to be **29·9 kJ mol^{-1}**. Reference to the data in Table 10.1 suggests

that this group is likely to be a histidine (only the unprotonated form being active). It is now known that a histidine is involved in the mechanism of action of the enzyme.

10. From a plot of product formed vs. time, it is seen that the formation of PNP is biphasic—there is a rapid formation of product (largely completed with 15 s) followed by a steady state rate, for which the rate constant is $0 \cdot 0095 \text{ s}^{-1}$ (average value of the two experiments). A rough estimate of the magnitude of the fast phase can be obtained by extrapolation to $t = 0$, and the value obtained is about $0 \cdot 73$ **mol** PNP (mol enzyme)$^{-1}$. This value may indicate the presence of some inactive enzyme in the preparation, but it should be noted that this type of extrapolation is only approximate, since the steady state rate is not negligible. The proposed mechanism for this reaction involves acylation of a serine group on the enzyme:

$$\text{Enz}-\text{OH} + \text{PNPA} \xrightarrow{\text{Fast}} \text{Enz}-\text{O}-\overset{\overset{\displaystyle O}{\|}}{C}-\text{CH}_3 + \text{PNP}$$

slow (steady-state rate) → acetate

Formation of the acylenzyme is fast and its breakdown to regenerate fresh enzyme is the slow step (steady-state rate).

Chapter 11

1. Using $E = h\nu$ and $\nu = c/\lambda$, we obtain the following energies.
   (a) $479 \cdot 7$ kJ mol^{-1}.
   (b) $24 \cdot 0$ kJ mol^{-1}.
   (c) $1 \cdot 2 \times 10^{-2}$ kJ mol^{-1}.
   (d) $2 \cdot 4 \times 10^{-5}$ kJ mol^{-1}.
   (e) $4 \cdot 85 \times 10^{-7}$ kJ mol^{-1}.

2. The *incident light* of wavelength 550 nm corresponds to an *energy* of $3 \cdot 6 \times 10^{-18}$ J per photon.
   The *minimum detectable energy* is 2×10^{-16} J s^{-1}. Thus, the *minimum rate of incidence of photons* is $(2 \times 10^{-16})/(3 \cdot 6 \times 10^{-18})$ s^{-1}, i.e. approx **550** photons s^{-1}.
   This illustrates the extreme sensitivity of the eye.

3. Using the equation

$$\text{Absorbance} = \varepsilon \text{ c.l.}$$

$$\varepsilon_{280} = 1 \cdot 11 \times 10^7 \text{ cm}^2 \text{ mol}^{-1}.$$

4. The NADH can be calculated from the absorbance at 340 nm. The total (NAD$^+$ + NADH) can be calculated from the absorbance at 260 nm.
   This gives

$$[\text{NAD}^+] = 13 \cdot 5 \text{ } \mu\text{mol dm}^{-3}$$

$$[\text{NADH}] = 33 \cdot 8 \text{ } \mu\text{mol dm}^{-3}$$

5. From a graph of extinction coefficient vs. pH, we can determine the pK_a from the

half point of the titration (see Chapter 7).
This gives

$$pK_a \text{ (nitrotyrosine)} = \textbf{6·8}$$

$$pK_a \text{ (nitrated enzyme)} = \textbf{7·9.}$$

The two values differ because the environment of the nitrotyrosine on the enzyme is different from that of the free nitrotyrosine. The increase in pK_a means that $\Delta G°$ for the ionization is more positive in the case of the nitrated enzyme (i.e. the unionized form is preferentially stabilized).

6. (a) The decrease in absorbance at 340 nm corresponds to the oxidation of 83·6 nmol NADH min^{-1} cm^{-3}. The amount of enzyme added is $0·02 \times 0·08$ mg. Thus the activity of the enzyme is **52·3** µmol substrate consumed min^{-1} mg^{-1}. **(0·87 katal kg^{-1}.)**
   Note that 1 mol of NADH oxidized is equivalent to 1 mol of ATP and creatine consumed in this coupled assay.

   (b) The NADH consumed is equivalent to 128·6µmol dm^{-3}. Thus the stock solution of creatine is $50 \times 128·6$ µmol $dm^{-3} = \textbf{6·43 mmol dm}^{-3}$.

   Note that the creatine kinase equilibrium, which would normally lie well over to the left (Chapter 2), is driven over to the right by coupling to the pyruvate kinase and lactate dehydrogenase systems.

7. At equilibrium the concentration of NADH formed $= 0·090$ mmol $dm^{-3} =$ $[NH_4^+] = [\alpha\text{-ketoglutarate}^{2-}]$. The concentrations of glutamate$^-$ and NAD$^+$ are obtained by difference. Noting that $[H^+] = 10^{-7}$ mol dm^{-3} and that H_2O is in its standard state, K is evaluated as **4·05 $\times 10^{-15}$ (mol dm^{-3}).**
   The amount of NADH formed is very small, compared with the concentrations of glutamate$^-$ and NAD$^+$. Since the concentration of NADH formed is raised to the third power in the expression for K, any inaccuracies in estimating this concentration constitute a major source of error in determining K. For accurate work, the effects of ionic strength on the activity coefficients (and thus on K) should also be considered. In practice, the true equilibrium constant would be obtained by an extrapolation to zero ionic strength (infinite dilution).

8. By constructing a standard curve (absorbance vs. amount of Hb) it can be seen that the 0·01 cm^3 of blood contains 1·4 mg Hb; i.e. the Hb content is **140 g dm^{-3}** (A doctor would express this as 14 g $(100 \text{ cm}^3)^{-1}$.) Normal ranges for males are 130–180 g dm^{-3} and for females 110–160 g dm^{-3}.
   Note that it is important to have readings from a range of standards above and below the sample reading, since linearity of absorbance vs. amount cannot necessarily be assumed.

Chapter 12

1. The amount of phenylalanine originally present was **11·1 g.**
   We derive this result by determining the amount of phenylalanine in the radio-active sample added. This is $0·2$ µmol (0·033 mg) and possesses 10 microcuries of activity (i.e. $2·22 \times 10^7$ d.p.m.). Since after dilution the activity of the phenylalanine isolated is 2000 d.p.m. mg^{-1} the dilution factor can be evaluated as $3·36 \times 10^5$ (i.e. the phenylalanine originally present is $3·36 \times 10^5$ times that added).

2. The rate constant for ^{32}P decay can be calculated to be 0.0488 day (i.e.3.389×10^{-5} min^{-1}).

   Now the rate of decay is given by $-dN/dt = kN$ (N = number of atoms present). From this N can be calculated as 2.87×10^7 atoms (i.e. **4.77×10^{-17} g** atoms P).

3. The tRNA added is 4×10^{-10} mol.

   By a similar calculation to Question 1, the amount of L-alanine bound to the tRNA is 3×10^{-10} mol.

   Hence **75** per cent of the tRNA is converted to L-alanine-tRNA.

4. The decay constant k is the first-order rate constant which characterizes radio-active decay. The *half life* of an isotope is the time for the number of radioactive atoms to fall to half its orginal value; k is given by $t_{\frac{1}{2}} = \ln 2/k$. (For more details see text.)

   The ln (radioactivity) is plotted against time to obtain an overall decay constant of 0.185 day^{-1}. This is made up of (i) the radioactivity decay of the isotope and (ii) the loss of the compound by turnover and excretion. Since the radioactive decay constant is 0.08 day^{-1} the 'biological' decay constant is 0.105 d^{-1}, i.e. the biological half life of X is **6.6** days.

Some useful constants and conversion factors

Planck constant	h	$6 \cdot 63 \times 10^{-34}$ J s
Speed of light in a vacuum	c	3×10^{8} m s^{-1}
Avogadro number	N_{A} (or L)	$6 \cdot 02 \times 10^{23}$ mol^{-1}
Gas constant	R	$8 \cdot 31$ J K^{-1} mol^{-1}
Faraday constant	F	$96\,500$ C mol^{-1}
Curie	Ci	$3 \cdot 7 \times 10^{10}$ s^{-1}
Svedberg	S	10^{-13} s
Molar volume of ideal gas at 273·15 K and 1 atm	$=$	$22 \cdot 41$ dm^{3} mol^{-1}
1 electron volt	$=$	$96 \cdot 5$ kJ mol^{-1}
1 calorie	$=$	$4 \cdot 18$ J
0°C (centigrade)	$=$	$273 \cdot 15$ K
1 atm (760 mm Hg)	$=$	$101 \cdot 325$ kPa
$\ln x$	$=$	$2 \cdot 303 \log_{10} x$

Some suggestions for further reading

Thermodynamics
I. M. KLOTZ: *Energy changes in biochemical reactions*. Academic Press, New York (1967).
A. L. LEHNINGER: *Bioenergetics*. W. A. Benjamin, New York (1965).
E. A. NEWSHOLME and C. START: *Regulation in metabolism*. John Wiley, London (1973).

More advanced texts
K. G. DENBIGH: *The thermodynamics of the steady state*. Methuen, London (1962).
C. TANFORD: *Physical chemistry of macromolecules*. John Wiley, New York (1961).
K. E. VAN HOLDE: *Physical biochemistry*. Prentice-Hall, New Jersey (1971).
H. GUTFREUND: *Enzymes: physical principles*. Wiley-Interscience, London (1972).

Acids and bases
R. G. BATES: *Determination of pH*. John Wiley, New York (1964).
R. P. BELL: *The proton in chemistry*. Cornell University Press, Ithaca, NY (1973).

Ultracentrifugation
H. K. SCHACHMAN: *Ultracentrifugation in biochemistry*. Academic Press, New York (1959).
C. TANFORD: *Physical chemistry of macromolecules*. John Wiley, New York (1961).
K. E. VAN HOLDE: *Physical biochemistry*. Prentice-Hall, New Jersey (1971).

Oxidation–reduction processes
W. J. MOORE: *Physical chemistry*. Longmans, London (1972).

Chemical kinetics
W. J. MOORE: *Physical chemistry*. Longmans, London (1972).

Enzyme kinetics
K. J. LAIDLER and P. S. BUNTING: *The chemical kinetics of enzyme action*. (2 edn). Clarendon Press, Oxford (1973).
H. GUTFREUND: *Enzymes: physical principles*. Wiley-Interscience, London (1972).
W. FERDINAND: *The enzyme molecule*. John Wiley, London (1976).
A. CORNISH-BOWDEN: *Principles of enzyme kinetics*. Butterworths, London (1976) and *Fundamentals of enzyme kinetics*. Butterworths, London (1979).
P. C. ENGEL: *Enzyme kinetics*. Chapman & Hall, London (1977).
A. FERSHT: *Enzyme structure and mechanism*. W. H. Freeman, Reading, Mass. (1977).

Index